DICTIONARY *of* ASTRONOMICAL NAMES

DICTIONARY *of* ASTRONOMICAL NAMES

Adrian Room

R

ROUTLEDGE
London and New York

He telleth the number of the stars;
he calleth them all by their names.

(Psalm 147: 4)

First published in 1988 by
Routledge
11 New Fetter Lane, London EC4P 4EE

Published in the USA by
Routledge
a division of Routledge, Chapman and Hall, Inc.
29 West 35th Street, New York, NY 10001

Set in Linotron Times
by Input Typesetting Ltd, London SW19 8DR
and printed in Great Britain
by
T. J. Press (Padstow) Ltd,
Padstow, Cornwall

Library of Congress Cataloging in Publication Data
Room, Adrian.
Dictionary of astronomical names/Adrian Room.
p.   cm.
Bibliography: p.
1. Astronomy—Dictionaries.   2. Astronomy—Names.   I. Title.
QB14.R66 1988
520′.3′21—dc19   87–31670

63,932

British Library CIP Data also available

ISBN 0–415–012988

CONTENTS

ACKNOWLEDGMENTS

I should like to acknowledge specialist assistance – essential for a work such as this – from two sources.

First, I should like to thank Patrick Moore for advising on suitable source books for the origin of astronomical names, in particular those of craters on the Moon and of the minor planets. I particularly wish to thank him for carefully reading the text of this book at typescript stage, and for saving me from the various snares and delusions that can occur – and in some cases have occurred – in works of this kind.

Second, I warmly thank the Royal Astronomical Society, London, in the persons of the Society's librarian and assistant librarian, Peter Hingley and Pamela Towlson respectively, for readily loaning valuable source books and for providing helpful photocopy material, as well as for making particular researches and enquiries on my behalf. This present Dictionary would be a much slimmer and poorer volume without their willing help.

I also wish to thank the Royal Astronomical Society for permission to reproduce the Map of the Moon on p. 168 from Riccioli's *Almagestum Novum* of 1651, a copy of which is held in the Society's library.

For background information about the origin and activities of the International Star Registry, referred to in the Introduction, I wish to thank Ann Herzberg, one of the organisation's officers.

INTRODUCTION

Ever since man evolved as *Homo sapiens*, millions of years ago, he has also been *Homo curiosus*, 'inquiring man', and one of the unceasing objects of wonder for him has been the universe around him, all that is 'up there' in the sky and beyond, all that is unknown and still largely uncharted territory.

There is much to marvel at, indeed. There is the daily Sun, with its heat and light, and the nightly Moon, with its associated cold and pale illumination. There are the stars, also nightly visitors, some apparently near and some far, some large and bright and others small and faint, some isolated and easily distinguishable and others fused in a blurred luminous mass, some seemingly emitting a steady light and others ceaselessly flickering and twinkling. (It was later that the 'untwinkling' stars were discovered to be different bodies, and were designated as planets.) Then there are the irregular visitors, the comets and meteors and seasonal visual phenomena, making an impressive and often spectacular appearance in the night sky, and serving to increase man's wonderment and curiosity.

All these remote but fascinating bodies were undoubtedly powerful and influential, too. Not only did they appear and process with predictable patterns, but also they clearly influenced much of the natural world in which man himself lived. The regular phenomena of day and night, light and darkness, summer and winter, high tide and low tide, growth and decay, was surely governed by these powerful heavenly creations, themselves subject to, and even part of, the supreme will of a divine guiding force.

Gradually, with the advance of science and a more sophisticated and objective reasoning, man came to realise that many of his earlier suppositions and attitudes about the nature of the universe had been incorrect. For a start – what a start it was! – man determined that his world was not, after all, the centre of the universe,

1

with all the other bodies revolving and wheeling round it, but that it was, in turn, one of a smallish number of similar bodies that itself revolved round the Sun. The Sun, too, was discovered to be a star, and the Moon found to have its equivalent for other planets that orbited the Sun.

It was therefore necessary to modify the terminology in many ways with regard to the night sky and its phenomena, and some of the original 'stars' are now called 'planets', while so-called 'shooting stars' are today mostly correctly referred to as 'meteors'.

More radically, it became necessary to distinguish between 'astronomy' on the one hand, and 'astrology' on the other. Formerly, 'astrology' was a term that embraced what is known as 'astronomy' today, and was itself subdivided into 'natural astrology', which concerned itself with calculating the movements of the heavens, and 'judicial astrology', which studied the supposed influences of 'heavenly bodies' on human life and destiny, much as 'astrology' continues to do in its modern form.

It was the stars, and the groupings of stars, that particularly awed and fascinated the ancient astrologers, especially when they were in some way different or remarkable from their fellows. It was obvious that names would have to be devised for such stars and star clusters, names that would somehow be appropriate for their appearance or 'performance' or supposed influence. The heavenly bodies that are called 'Sun' and 'Moon', in whatever language, are clearly very different from other celestial creations, and although the Moon itself later came to acquire its own nomenclature for the different features that could be observed on it, it was first and foremost the stars that needed names. From the earliest times, too, the stars were found to have a practical value as well as a supposed influential one, because when travelling, to this place or that, people could 'follow a star' in order to go in the required direction. Even today, in the sophisticated space age, stars are still important for direction-finding for travellers of all kinds, whether their medium is sea, air or the cosmos. So here was another good reason to name the stars!

All round the world, in different lands, people devised different names in different tongues for what they saw in the night sky. (Recent studies have shown, fascinatingly, that some of the names given by quite unrelated peoples were essentially the same, with Chinese star names, for example, identical to those given by the native Indians of the American continent.)

The oldest astronomical (or astrological) names, and the ones that are still familiar to many today, are those of the signs of the zodiac, which are not only designations for particular belts of the heavens, but also remain as names for the constellations originally in those belts, still picturesquely evoking the outline of the figure observed there. (At one time, the signs actually corresponded to the constellations in front of which the Sun appeared to pass in its yearly path round the heavens. Today, however, the signs of the zodiac and the constellations, although sharing the same name, are quite distinct, and now the sign of the zodiac called Aries, for example, is located among the stars of the constellation of Pisces. For a modern astrologer to talk of the 'Sun entering Aries' on a particular date is thus quite different from an astronomer speaking of its entering the constellation of the name on a certain date.)

The constellation names in fact came first, and were devised, it is believed, some time in perhaps the fourth millennium BC by the Sumerians, a Middle Eastern people. They could see the pictorial outline of a ram in what became the constellation of Aries, for instance, and an archer with his bow in what later became the constellation of Sagittarius. Much later, the Ancient Greeks translated the names and evolved an arrangement that divided the heavens into twelve belts, each named after a constellation and occupying one-twelfth (that is, 30°) of the so called 'great circle'. The Greeks called their scheme *zodiakos kyklos*, literally 'circle of animals', because with one exception each band contained a constellation that was itself named after an animal or a living creature. (There are seven 'animal' names proper, corresponding to English Ram, Bull, Crab, Lion, Scorpion, Goat and Fishes; the other four are human figures, known in English as the Twins, the Virgin, the Archer and the Water Carrier.)

The only exception here is Libra, the Balance (or the Scales). How did it come about that eleven of the signs of the zodiac have 'animal' names, but only one a 'non-animal' name? And why a Balance? The answer is a straightforward one. The ancient astrologers noticed that when the Sun passed before this particular group of stars in the sky, it did so at the time of the year when the day was equal in length to the night. In other words, it did so at what we now call the autumnal equinox. Hence the 'Balance' that symbolised the equal value in time of night and day. (For an explanation of this and similar technical terms, see the Astronomical Glossary on pp. 31–47.)

The Greek names were then translated by the Romans in the Latin forms familiar to us today. Later still, each sign of the zodiac

3

(or its constellation name) was assigned a particular symbol, such as the two parallel wavy lines for Aquarius. These appear to have arisen some time in the Middle Ages. For an explanation of their possible interpretation, see the Glossary.

The actual size and number of the zodiacal constellations, incidentally, originally varied considerably, and it was the Greeks who established the number as twelve (that is, twelve belts of 30° each to give the great circle total of 360°) and thus determined their names as these particular ones. Ptolemy, in the second century AD, recognised forty-eight constellations, and since his time further constellation names have been added, making a fixed total today of eighty-eight. We shall consider these again in due course.

The peoples of the Middle East, in the earliest days, also named individual stars, and many of these are still popularly known today in their Arabic form, or at least in a version that to some degree represents the original Arabic. Most of us will have heard or read of Vega, Betelgeuse and Aldebaran, for example, if only in science fiction or 'space adventure' tales or films. However, by no means all stars names are Arabic, and very many are Greek or Latin, so that we now have something of a pot-pourri of astronyms to savour and remember. Among well-known Greek names, for example, are Sirius, Procyon and Arcturus, and Latin names include Polaris, Capella and Regulus.

Another familiar star name to many is Proxima Centauri, which is a mixture of Latin and Greek, meaning 'nearest one of Centaurus', the latter word being the name of the constellation, and a reminder that although the majority of current constellation names are Latin, many are Greek in origin, and relate to characters and incidents in classical mythology. (Other Greek constellation names are Boötes, Hydra, Monoceros and Ophiuchus. It is fair to say, though, that the Romans simply adopted these in Latinised form rather than translating them.)

The name of Proxima Centauri is also a reminder that very many stars, even the well-known Sirius and Polaris, have an alternative 'two-part' name like this, with the first word a letter of the Greek alphabet, and the second word the name of its constellation with the Latin genitive ending. 'Centauri' (meaning of 'Centaurus') is thus the second word in the alternative designation of all named stars in that constellation, so that its brightest star, named as Rigil Kentaurus (which happens to be half-Arabic and half-Greek) is also Alpha Centauri. The latter type of name is the one that astronomers today mostly employ, and it is true to say that in many cases the old Arabic or Greek or Latin names are used less and less, except for popular reference, or by latter-day namers who are looking for a suitable name with which to name something, whether a Sirius radio-microphone or a Polaris submarine. Moreover, the star that astronomers know usually as Beta Centauri, still appears in some astronomical maps and books by one or other of its *two* original names of Hadar or Agena, making a total of three different names. It can be seen, therefore, that astronomical names can be complex affairs, involving at least three ancient languages (Arabic, Greek and Latin)!

The links between the stories of classical mythology and the names of astronomy are easier to see in many of the non-zodiacal constellation names than in those that correspond with the signs of the zodiac. In fact, many constellation names are those of well-known mythological characters, such as Hercules, Perseus, Pegasus, Orion and Andromeda. All these are old names, and were among those recognised by Ptolemy. Names of individual stars that are those of mythological characters are much smaller in number, however, with the best known being those of the classical twins, Castor and Pollux. Even so, most of the old constellation names, even where not the same as those of the mythological personages, do connect fairly closely with the classical stories, and in a way are almost a 'visual aid', or at least a memory refresher, to the events contained in those stories. This is particularly true of the northern hemisphere, where the names are generally much more ancient than a large number of those of the southern skies. Their precise connections are told in the entries that follow in this Dictionary, but suffice it to say, for instance, that Aquila (the Eagle) represents the bird that accompanied Jupiter on his travels, that Sagitta was the Arrow shot by Hercules, and that in the southern hemisphere, Lepus, the Hare, lies at the feet of Orion, the mighty hunter. It is particularly pleasant when the names of two adjacent constellations combine to give an even more detailed and precise mythological picture. (Sagitta, the Arrow, for example, lies between Aquila, the Eagle, and Cygnus, the Swan, while also adjoining it is Hercules himself, the great classical hero who shot it, although some identify it with Cupid's bow.)

Apart from the ancient names of individual stars and constellations, some of the names of star formations in the night sky are equally historic, and equally famous. One of the best-known star patterns, for example, is that of the Plough, whose seven components form part of the constellation of Ursa Major. Not all countries have seen the formation as that of a plough, however, and in many parts of Europe the depiction was seen as that of a cart or chariot, while in the Middle East the outline was seen to be that of a coffin. (The first of these images gave the alternative name of Charles's Wain.) Across the Atlantic, meanwhile, a common American name for the seven stars is the Big Dipper, that is, as a ladle with a long handle. Elsewhere, in other lands and by other tongues, the starry septet has been thought of as seven individuals of some kind, who have been designated by such collective names as the Seven Shiners, the Seven Sages, the Seven Bulls, the Seven Sleepers of Ephesus, the Seven Champions of Christendom, the Seven Little Indians, and indeed by almost any group name that features the magic and mystic number 'seven'. Alas, it has not proved practicable in this Dictionary to list the many alternative names that have been recorded for some of the best-known features of the sky.

Again, other familiar formation names are those of the Milky Way and the Galaxy, and these two names are in turn examples of the way in which certain ancient names have linguistic and lexical links, as well as narrative and pictorial ones. The name 'Milky Way' is an English translation, for instance, of Latin 'Via Lactea', which in turn is a translation of Greek 'Kyklos Galaktikos' (meaning 'milky circle'), based on the same Greek word for 'milk', *gala*, genitive *galaktos*, that gave not only Latin *lac*, genitive *lactis*, but also English 'galaxy'. 'Milk' thus runs through both names and both classical words. In other cases, an echo of one constellation name can be found in another, such as Sagitta and Sagittarius, Taurus and Centaurus (although one was all bull, and the other half horse), Serpens (from Latin *serpens*, 'snake') and Ophiuchus (from Greek *ophis*, 'snake'), Hydra and Hydrus, Canis Major (and Canis Minor) and Canes Venatici. However, in the case of the last two couples here, it is only the first name that is ancient, and the second name is that of a constellation added only in the seventeenth century. Even so, the linguistic link is there, perhaps even confusingly in some instances.

It should not be assumed, incidentally, that the Arabic star names were the originals. The Arabians themselves translated many names that had arisen in the Middle East before them. Many of these, as mentioned, have remained today in their Arabic form, while others have been first translated by the Greeks then translated by the Romans. In all instances, such Arabic names are included in the Dictionary under the respective entries for these names. To what extent the Greeks actually originated any names, as against adopting them from 'ready-made' sources, is a matter that has still not been fully resolved even today by classical scholars. The originality of the Greek names ties in closely with the originality of the Ancient Greek myths, many of which are known to have been adopted from elsewhere, with the characters bearing different names in older languages. The whole subject is a complex one, and the identity of Aphrodite, for example, with an Asiatic goddess similar to Ishtar, the Babylonian goddess of love and war, who has been herself identified with the Canaanite Astarte, the Israelite Ashtoreth and the Arab god Athtar, is both involved and intricate. But the fact remains that it was the Greeks who made the constellation names familiar to the peoples of other lands, if only through Ptolemy's listing of forty-eight of them. Apart from the twelve zodiacal names, he recognised twenty-one names in the northern celestial hemisphere, and fifteen in the southern. These were respectively: (northern hemisphere) Ursa Major, Ursa Minor, Draco, Cepheus, Boötes, Corona Borealis, Hercules, Lyra, Cygnus, Cassiopeia, Perseus, Auriga, Ophiuchus, Serpens, Sagitta, Aquila, Delphinus, Equuleus, Pegasus, Andromeda, Triangulum; (southern) Cetus, Orion, Eridanus, Lepus, Canis Major, Canis Minor, Argo Navis, Hydra, Crater, Corvus, Centaurus, Lupus, Ara, Corona Australis, Piscis Australis. All these still exist in modern star maps and charts, with the exception of Argo Navis, which has now been subdivided into four smaller constellations. (See its entry for details of these.)

In speaking of ancient astronomical names, however, we have so far said nothing of a smaller but just as important group, whose own names are probably more familiar than any others. These are the names of the planets, those celestial bodies that, like our own Earth, comprise with their attendant satellites the Solar System, and that revolve round the Sun in ever-increasing and uniquely varying orbits. There are – at least, at present – nine known planets, and apart from the Earth their names, in order from the Sun, are Mercury, Venus, Mars, Jupiter, Saturn, Uranus, Neptune and Pluto. All names are well known from mythology, and all are familiar gods (and one goddess) with particular attributes or 'kingdoms'. At first glance, too, they all seem to be the names of Roman deities who have Greek equivalents. Thus Mercury corresponds to Hermes, Venus to Aphrodite, Mars to Ares, Jupiter to Zeus, Saturn to Cronos, Neptune to Poseidon, and Pluto to. . . To whom? He is the exception, because Pluto, the god of the underworld, was a

Greek god with no Roman equivalent. (The equivalent is not Hades, which is simply an alternative name.) But that is not all, for only the first five names are ancient, as only Mercury, Venus, Mars, Jupiter and Saturn are visible to the naked eye!

The fact of the matter is that the ancient astronomers (or astrologers) regarded seven heavenly bodies as 'planets'. These were (in the order of their accepted distance from the Earth) the Moon, Mercury, Venus, the Sun, Mars, Jupiter and Saturn. Hence the use of the phrase 'the seven planets' (magic 'seven' again) in medieval and even later writings. The five planets proper, however, were not originally known by their present names but by Greek names, and in Pythagoras' day (the sixth century BC) Mercury was called Stilbon (meaning 'shining one'), Venus was known as either Hesperos ('evening one') or Phosphoros ('light-bearing one') or Eosphoros ('bringer of the dawn'), Mars was Pyroeis ('fiery one'), Jupiter was Phaethon ('glittering one'), and Saturn was Phainon ('shining one'). Venus had different names because the planet could be seen both in the evening (as Hesperos) and in the morning (as Phosphoros or Eosphoros), and was generally believed to be two distinct celestial bodies. (It was Pythagoras, in fact, who first proposed that these were actually one and the same body.)

It will be noticed that all these names are connected with light or heat (or both). Mercury was a 'shiner' or 'sparkler' because it always accompanied the Sun, and was regarded as a sort of 'spark' flying off it. Mars was 'fiery' because of its red colour, and Jupiter was 'glittery' because it was noticeably bright at night, when the even brighter Venus was not visible.

It was then the turn of Aristotle, in the fourth century BC, to introduce personal mythological names for the five planets, and he thus renamed Mercury as Hermes, Mars as Ares, Jupiter as Zeus and Saturn as Cronos. (To be precise, he referred to the five by terms translating literally as 'Hermes' star', 'Zeus's star' and so on, and it was only subsequently that the 'star' or planet was known by its deity name alone.) Then, when the Greeks realised that Venus was one and the same planet, they came to call it Aphrodite, and this name also occurs in the writings of Aristotle, as well as those of his teacher Plato.

Even the personal names carried something of a descriptive aura. Mercury, for example, is the 'fastest' planet, as the one with the shortest sidereal period, and Mercury (or Hermes) was the swift-footed messenger of the gods. Mars (or Ares) was the god of war, and so had a red colour that was apt for this role. Jupiter (or Zeus) was given the name of the chief of the gods because the planet is the brightest in the sky (apart from Venus, which, as mentioned, is not visible at night when Jupiter is and, occasionally, Mars, which can sometimes outshine it). Saturn (or Cronos) follows Jupiter in order, as is appropriate for the god who in mythology was the father of Zeus. And Venus (Aphrodite) bore the name of the goddess of love, as was fitting for a heavenly body that appeared in the freshness of the morning or the romantic twilight of the evening. (Compare 'aubade' and 'serenade' for related concepts, as the terms for a romantic poem or song sung at either dawn or dusk respectively.)

The Greek names then passed to Ancient Rome, where they were translated by the names of those Roman deities who corresponded to the Greek, and these are the names that remain with us today for the planets, in many of the world's languages. For a period, though, the Romans believed Venus to be two separate bodies, as the Greeks had done, and so called the planet by two names, a 'morning' one and an 'evening' one. These were respectively Lucifer ('light-bringer') and Vesper ('evening one'). The final stage in planet-naming then followed considerably later, in modern times, although even then the names chosen for the three newly discovered planets, Uranus, Neptune and Pluto, were given in the same tradition and in the same manner as the ancient names. For their stories, and for further details on the names of the five original planets, see the appropriate entries in the Dictionary.

Astronomical names thus broadly divide into 'ancient' and 'modern', with the former category applying to the star, planet and other celestial group names that we have already considered. All other names are 'modern', which effectively means mostly seventeenth century or later, and applies in particular to the more recent constellation names of the southern skies and the names of the three outer planets in the Solar System just mentioned. 'Modern', too, is the category of name that must generally be used of two special types of astronomical name: those of the asteroids or minor planets, and those of features (such as craters) on the surface of the Moon. Of these two sizeable subgroups, the Moon names are generally the older, so as we are proceeding chronologically, we should consider them first.

Anyone who has looked at the Moon with just the naked eye will have observed that its surface has an irregular or 'blotchy' appearance, with some areas brighter and others darker. Fanciful Moon-gazers have discerned a 'man in the moon' as a sort of composite figure formed from the surface features. (The most popular outline conjured up seems to have been that of a man leaning on a pitchfork, with this interpretation perhaps additionally prompted by the biblical tale about the man who 'gathered sticks upon the sabbath day' told in Numbers 15: 32–6.) But other observers of old saw the Moon, quite literally, as a sort of mirror image, albeit a pale and small one, of the Earth, and as having the same kind of natural features as its inhabited counterpart, that is, seas, plains, mountains and valleys. After all, what else could the variegated surface markings be? When Galileo perfected the already existing imperfect telescope and discovered craters on the Moon, and sketched what he saw through his new refracting lens, it seemed more likely than ever that the Moon's surface was similar in many ways to that of the Earth. And such features would clearly need to be named. . . .

However, although Galileo, in the sixteenth and seventeenth centuries, may have been the first selenographer, or Moon studier (in effect a sort of 'lunar geographer'), he was not the first selenonymist, or Moon-namer. That distinction went to the seventeenth-century Belgian astronomer, Michel Florent van Langren, who was sent by Isabella, Regent of the Netherlands, to Spain, where Philip IV appointed him Court Astronomer. On his map of the Moon, published in 1645, Langren introduced some three hundred names for the most prominent features, with the names themselves those of famous real or fictional personages, from biblical characters and saints to members of Philip IV's family and holders of high office in his court. These first names did not on the whole survive, however, and today only three still appear on modern lunar maps: the three craters Catharina (for St Catherine), Cyrillus (for St Cyril) and Theophilus (for St Theophilus). All three are close to one another in location.

Two years later, in 1647, the German astronomer Hevelius published his famous *Selenographia*, or atlas of the Moon, with detailed maps of the surface and entirely new names. He did not follow Langren's 'biographical' principle but instead transferred geographical names from the Earth, especially the old, historic ones. Thus his map had a 'Euxinus Pontus', or 'Euxine Sea', this being the ancient name of the Black Sea, and the present craters named Copernicus, Tycho, Thales and Endymion were named by him respectively as 'Sicilia Insula', 'Sina Mons', 'Sarmatici Montes' and 'Lacus Hyperboreus' otherwise Sicily, Mount Sinai, Carpathian Mountains and 'Northern Lake' (that is, Arctic Sea). Most of his names have since fallen into disuse, as Langren's have, but like Langren's, one or two remain, among them the lunar mountain ranges still called the Alps and the Apennines. But although Langren had one or two 'watery' names on his earlier map, such as the 'Mare Astrologorum' or 'Sea of Astronomers' (today's Mare Frigoris, or 'Sea of Cold'), it was Hevelius who really put, quite literally, such names on the map. Before him, Leonardo da Vinci had suggested that the lighter-coloured markings on the Moon could represent areas of water, but now Hevelius gave areas – the darker-coloured ones, however – specific hydronyms or 'water names', using Latin words such as Oceanus ('ocean'), Mare ('sea'), Lacus ('lake'), Palus ('marsh') and Sinus ('strait'). This despite the fact that there is no water on the Moon! That such names have nevertheless become officially established can be shown by the fact that areas of the Far Side of the Moon, discovered only in the second half of the twentieth century, have also been given the names of 'seas', such as the Mare Ingenii (omitted from many modern maps) and Mare Moscoviense.

But it was the third 'Moon-namer' whose names were to survive in the greatest numbers. He was the Italian astronomer Francesco Grimaldi, whose map of the Moon, containing about three hundred names, was published in Naples in 1651, four years after Hevelius's own map.

Grimaldi was a student of the better-known Italian astronomer Giovanni Riccioli, and his map was published by Riccioli as part of the latter's work entitled *Almagestum Novum* or 'New Almagest', the second word deriving from the Arabic meaning 'the greatest', but given directly as a tribute to Ptolemy's great treatise on astronomy known as the *Almagest*. As it was Riccioli who was the overall author of the work, he usually gets the credit today for the Moon names that were actually introduced by Grimaldi, which is rather unfair. Be that as it may, it was these names that became established, and today over two hundred of them are still in use for lunar features. Mare Imbrium, Mare Crisium, Mare Nectaris, Mare Nubium, Mare Vaporum, Mare Tranquillitatis (the latter to become famous with the first Moon landing of Apollo 11 in 1969) – all these were names that first appeared on Riccioli's (Grimaldi's) map.

Not all of Riccioli's generic names have been retained, however. For example he (or Grimaldi) introduced a number of names called 'terrae', or 'lands', but these no longer exist, and his original 'Terra Grandinis' ('Land of Hail'), 'Terra Mannae' ('Land of Manna'),

Introduction

and great 'Terra Caloris' ('Land of Heat') and 'Terra Sanitatis' ('Land of Health') have long since been expunged from the map, as have his 'Terra Vitae' ('Land of Life') and 'Terra Siccitatis' ('Land of Dryness'). Many of these names seem to have been designed to have an 'aqueous opposite', that is, a 'sea' name that contrasted with the 'land' one. These remain, such as the Lacus Mortis, or 'Lake of Death' to contrast with the now vanished 'Terra Vitae', and the Mare Frigoris, or 'Sea of Cold', as opposed to the original 'Terra Caloris'. But the effect of such dramatic contrasts is now lost, because one half of the pair is missing.

As to *why* there are no longer any 'Terrae' on the Moon, there are two good reasons. One is that the 'lands' were not really distinctive in the way that the darker-coloured 'seas' are, and they were in some cases almost too extensive in area. Second, the Latin word 'terra' gives the word in some languages that also translates as 'earth' (for example in French), and this is the name of our own planet! To have 'Earth' names on the Moon is thus inappropriate or even confusing.

Riccioli, like Langren, named a crater after himself, a large walled plain near the western limb (edge) of the moon. But because he did not believe in the Copernican theory, that the Earth revolves round the Sun, he showed his disapproval when he named a crater after him, and thus 'flung Copernicus into the Ocean of Storms'. And today it is in the Oceanus Procellarum that the prominent crater Copernicus remains.

In his naming of craters, Riccioli (that is, Grimaldi) made a major and orderly contribution to lunar nomenclature, unlike the earlier, more random system followed by Langren. Riccioli's names are biographical, like Langren's, and honour the famous, but in an altogether more appropriate and thoughtful way. For the craters of the Moon's northern hemisphere, he introduced the names of famous ancient philosophers and men of learning, especially astronomers, keeping the 'big' names for the larger craters, such as Aristotle, Archimedes, Aristarchus, Herodotus, Pythagoras, Plato and Thales. This inevitably involved considerable renaming, and Langren's craters named after St Athanasius, St Margaret and St Anthony, for example, now became respectively Plato, Ptolemy and Pliny. Nor did Riccioli (Grimaldi) overlook his immediate pioneering predecessors, and he assigned the names of both Langren and Hevelius to two craters.

Meanwhile he reserved the names of more recent scholars, those of the Renaissance, for the southern hemisphere, and it is here that

12

Tycho Brahe, Regiomontanus and (as already mentioned) Copernicus can be found. In some instances, he even contrived to give the names of the pupils of these great men to smaller craters close to their larger 'parent', so that they continue to sit, as it were, at the feet of their respective masters. Thus Rhaeticus, who was an associate of Copernicus, has his name in a small crater near the latter larger one. The same principle was even extended to distinguished writers for their literary 'offspring'. Thus *Timaeus* is the name of a famous dialogue by Plato, and on the Moon the former name is that of a small crater in the same region as the latter, prominent one (but on the other side of the Mare Frigoris). Even those two legendary strong men, Hercules and Atlas, are neighbours.

Knowledge of the particular evolution of such names, and of the men who gave them, today gives added point to the location of the craters named for Riccioli, Grimaldi and Langren. The first two are close together (master and pupil, after all) on the western edge of the Moon's disc, where not far from them can also be found Hevelius and Galileo. Langrenus (Langren), however, is located in isolation over near the eastern edge. This extreme disposition could be taken to imply Riccioli's (Grimaldi's) radical departure from Langren's system of nomenclature on the one hand, but on the other could also be symbolically regarded as the two Moon-namers 'embracing' the entire disc of the surface of the Moon, with their names lying across the surface between them.

In broad outline, these men established the pattern for lunar nomenclature that has been observed ever since, even for features on the Far Side, as we have seen. And it was not long before further geographical names followed, too, so that the original Alps and Apennines were joined by the Carpathians, Pyrenees, Caucasus and many more. True, there have been some deviations from the original style of 'sea' names, and the more recent names of *maria* have been either 'positional' (such as Mare Orientale, or 'Eastern Sea') or 'possessional' (such as Mare Humboldtianum, 'Humboldt's Sea', or Mare Smythii, 'Smyth's Sea').

The science of astronomy in general, and selenography in particular, developed fairly rapidly after the publication of the 'New Almagest', especially with the introduction of bigger and better telescopes and other instruments. But it was only in 1791, 140 years after Riccioli's publication, that any significant stride forward was made in Moon-mapping and Moon-naming. The advance was made thanks to the work of an amateur astronomer from Lilienthal, near

Bremen, where Johann Schröter, the chief magistrate, made many drawings of lunar formations and added more than seventy new named craters. (Alas, his private observatory was destroyed by invading French troops in 1813, and many of his valuable notebooks were lost.)

After Schröter, it was the two German astronomers Mädler and Beer who made the next significant contribution, when they published their great map of the Moon in Berlin in 1837. This added yet more names to the Moon's surface, some 140 in all, mostly introduced by Mädler.

It was Schröter, however, who had originally devised the method of designating a so far unnamed object by the name of the nearest (named) crater, plus an additional letter or number to indicate its particular nature. For example a depression, such as a crater or valley, would add a capital Roman letter (A, B, C, D, etc), an eminence, such as a peak or hill, would add a small Greek letter (α, β, γ, δ, etc), while rilles (*rimae*), observed as winding cracks or fissures, would add a Roman number (I, II, III, etc.) together with the letter 'r' (for *rima*, Latin for 'rille'). This system has been modified and improved in recent years, especially for designations of features on the Far Side of the Moon, but basically Schröter's method still applies.

Further lunar names have been added in subsequent years in more or less the same tradition, not least when the Soviet Moon probe, Luna 3, went right round the Moon in 1959 and took the first photographs of the Far Side, necessitating the urgent addition of dozens of new names. But they, too, followed in the same tradition, even though it was not philosophers and theologians who were now prominent, but scientists and explorers, 'pioneers' of relatively modern times. Such names were duly approved by the International Astronomical Union, which is today responsible for new nomenclature of all kinds in astronomy, and the range of current Moon names may be found in Appendix I, p. 169, where a comprehensive listing of named craters and other features is given, together with the origins of the particular names and an introductory account of the more recent, twentieth-century namings. Many significant or important Moon names also have their own entries in the main body of the Dictionary.

But we have reached the twentieth century in our voyage to the Moon, and we must now return for a time to the early nineteenth, or to be exact the final years of the eighteenth, when a search, ultimately successful, was made for entirely new astronomical objects to name.

It began with six astronomers meeting at Schröter's private observatory in Lilienthal in 1800. They were keen to put Bode's Law, as it was called, to the test.

The German astronomer Johannes Bode had drawn attention in 1772 to a peculiar 'law' of planetary distances. (The law had actually been formulated earlier that year by his fellow astronomer Johann Titius of Wittenberg, so it should more correctly be known as Titius-Bode's Law. Its popular name is nevertheless simply 'Bode's Law'.) The 'law' worked on the following simple mathematical principal. If you start with 3, and double each number successively (getting 6, 12, 24, 48, 96, 192 and 384), then add 4 to each figure, you get the approximate distances of the planets from the Sun if you regard the Earth's distance from it (= 1 astronomical unit) as 10 and add 4 to an initial 0 to get the distance of Mercury, the first planet. When Bode formulated his rule, Uranus, Neptune and Pluto had not yet been discovered, so his table looked like this (with the actual distances of the planets alongside in the same units for comparison):

Planet	Distance according to Bode	Actual distance
Mercury	4	3.9
Venus	7	7.2
Earth	10	10.0 (= 1 a.u.)
Mars	16	15.2
?	28	?
Jupiter	52	52.0
Saturn	100	95.4
?	196	?
?	388	?

This suggested not only that there could be one or more planets at roughly these distances beyond Saturn, but also that there even appeared to be a planet missing between Mars and Jupiter. Excitingly, when Uranus was discovered in 1781, its figures were shown to fit in well, as they were:

Uranus	196	191.8

Alas, much later, when Neptune was discovered (in 1846), its actual distance of 300.7 nowhere near corresponded to the calculated figure of 388, although interestingly, when later still Pluto was discovered (in 1930), it was found to have, on Bode's scale, a distance from the Sun of 394.6, which more or less did accord with the projected figure of 388.

But, meanwhile, what about that gap between Mars and Jupiter? Because the lower figures seemed reasonably reliable, that made it more probable that an important discovery awaited the world, and that somehow one of the planets in the Solar System had gone undetected.

The six astronomers, with Schröter as their president, therefore set up an organisation to search for the missing planet. They called themselves the 'Celestial Police', a name that today smacks more of science fiction and rock groups. But their purpose was serious enough, and then fortuitously, on 1 January 1801 (the first day of the first month of the first year of a new century – what an omen!), the great discovery was made. It was made, however, not by a member of the 'Police' but by the Italian astronomer Giovanni Piazzi, during the course of compiling a star catalogue. (Piazzi himself joined the 'Police' subsequently.) Moreover, it was not a new planet that he discovered, but a small, planet-like object. This was to be the first of the minor planets, or asteroids, subsequently to be named Ceres. (See *asteroid* in the Glossary, p. 31, for the history of the name.) From then on, more and more minor planets were discovered, but only slowly at first, and just four had been discovered by the time the 'Police' disbanded in 1815, and the fifth was not located until 1845.

Even so, a new question arose. What should these newly dis-covered bodies be named? What type of naming system, if any, should be followed? They were, after all, planets of a kind, and were situated in a planetary orbit in the Solar System. It seemed clear, therefore, that at least the first discoveries should have classical names, like those of the 'real' planets. This was actually the principle adopted, and the first minor planets were accordingly named Ceres, Pallas, Juno, Vesta, Astraea, Hebe, Iris, Flora, Metis and Hygeia, the last named being the tenth discovery, made in 1849. But after this, discoveries began to be made thick and fast, and it seemed that classical or mythological names could not be assigned indefinitely. They might even run out! Recourse was there-fore necessary to names of other types, such as those of distinguished human beings, and even geographical names. For details of these, and of what happened subsequently and is still happening (nearly 3000 minor planets have been named to date), Appendix II, p. 233, gives the names of the first 1000 minor planets to be discovered, together with brief accounts of their origins (where known); the names are examined thematically as they have been allocated over the years.

We have already considered the ancient names of the constellations, and have mentioned (and listed) the forty-eight names, including the zodiacal ones, recognised by Ptolemy in the second century AD. Something now needs to be said about the more recent names, especially of constellations in the southern celestial hemisphere.

Ptolemy's names, and no more, were in use down to the sixteenth century. They were limited to forty-eight because until that time astronomical observations had been restricted to the stars and constellations of the northern hemisphere (only). With the expansion of trade to the southern seas in the sixteenth to eighteenth centuries, and with the growth and development of round-the-world navigation, whether for exploration, discovery or territorial acquisition, new stars and constellations in the southern skies came into view for the first time. Clearly they would need names.

One of the best known and most distinctive constellations of the southern hemisphere is the Southern Cross, and it was known to have had this name as early as 1520, during Magellan's circumnavigation of the globe. This section of the sky was in fact also known to the Ancient Greeks, who had incorporated it in their own constellation of Centaurus. Then in about 1590 it is known that the great Danish astronomer Tycho Brahe had given the southern constellation of Coma Berenices its name. But now, after all the animals and mythological personages of the northern constellations, what kind of names would these new groups acquire?

It was the German astronomer Johann Bayer who set the first trend, when in 1603 he gave exotic 'Southern Seas' names to nearly a dozen constellations. He called them after rare and colourful birds and beasts, with names translating as Peacock, Toucan, Crane, Chameleon, Flying Fish and Water Snake, for example. In a sense he was following the tradition of the ancients here, inasmuch as his names were those of creatures. But their exotic nature was specifically chosen for their southern location, making them both traditional and yet original.

Hevelius was the next constellation-namer of any magnitude, and continued the 'animal' tradition, although without any flavour of the exotic. However, his names were *not* in the southern hemisphere, but the northern, where he aimed to fill unnamed gaps and formations with new names. Among the half dozen or so names of his which remain in use today are those translating as The Hunting Dogs, The Lesser Lion, The Fox, The Lizard and The Shield (originally Sobieski's Shield). Nothing too original or innovatory here. But Hevelius also used a reference to his favourite astronomical instrument as a name, and called one constellation The Sextant.

This last name seems to have been a pointer towards things to come, because when one of the most prolific constellation-namers, the French astronomer Nicolas Louis de Lacaille, introduced his own names for southern constellations in the mid-eighteenth century, they were almost all of scientific instruments and in particular those used in astronomy. Fourteen of his names still exist in use, and they include The Pendulum Clock, The Airpump, The Octant, The Compasses, The Level, The Telescope (of course!) and The Microscope, as well as the more oblique Sculptor's Workshop. One of Lacaille's names was The Table Mountain, denoting the geographical location of the observatory on the Cape of Good Hope, in South Africa.

Many of the names proposed for the newly discovered constellations were never accepted, and some eighteenth-century rejections include The Reindeer, Poniatowski's Bull, George's Lute, Herschel's Telescope, The Mural Quadrant, The Printing Press, and (doubtless because he had been sadly overlooked until now) The Cat. Some of these names have their own entries in this Dictionary, but mostly they are excluded, as they now are from modern star charts.

A word needs to be stated finally about the usage and status of astronomical names today. As already mentioned, astronomers tend to use the ancient names of stars only infrequently, preferring instead the two-part designation consisting of Greek letter plus the Latin name of the appropriate constellation in the genitive, such as Alpha Pegasi for one of the brightest stars in Pegasus, not Markab, its Arabic name. Such designations are usually abbreviated, with simply the Greek letter and a short form of the constellation name. This particular star would thus be referred to as α Peg. It is also a general principle that the brightest star is the one named Alpha, the second brightest Beta, and so on. But there are several prominent exceptions to this rule. For instance, in Orion and Gemini, it is the Beta star that is the brightest, and as a result of the subdivision of the ancient large constellation of Argo Navis into smaller constellations, neither Vela nor Puppis have Alpha or Beta stars at all. Moreover, a few constellations have gaps in the run of Greek letters, notably Carina, which has no Gamma or Delta, and also omits other letters.

For the purposes of reference, and to help with the order, here is the complete Greek alphabet with the name of each letter.

α	alpha	ι	iota	ρ	rho
β	beta	κ	kappa	σ	sigma
γ	gamma	λ	lambda	τ	tau
δ	delta	μ	mu	υ	upsilon
ε	epsilon	ν	nu	φ	phi
ζ	zeta	ξ	xi	χ	chi
η	eta	ο	omicron	ψ	psi
θ	theta	π	pi	ω	omega

In the Dictionary, however, such designations are always spelt out in full.

It will be noticed, in this book or elsewhere, that some stars have different designations, with for example a Roman letter, not a Greek one. Such stars will normally not have an ancient (Arabic, Greek or Latin) name, but may have a modern English name, which is really more a descriptive nickname. An example is the Blaze Star, in the constellation of Corona Borealis. Its official designation is T Coronae Borealis. Such letters often apply to variable stars (see Glossary), and originated when in the nineteenth century the German astronomer Friedrich Argelander assigned the Roman letters R to Z to conspicuous and unnamed variable stars in each constellation. After Z he then adopted the double form RR to RZ, then SS to SZ, TT to TZ and so on, until ZZ was reached. After this, one must start again at AA to AZ, BB to BZ, and so on, ending with QZ. This allows for 334 variables in each constellation, and if there are more than *that*, one simply has the letter V (for 'variable') and counts from 334, thus having V.335, V.336, and so forth. It may be asked why Argelander began his classification at R instead of A. This was because earlier letters in the alphabet had already been assigned to stars in some constellations (especially in the southern hemisphere) where the Greek letters had all been used up (to omega). One constellation already went down to Q, so he was obliged to commence his 'variable' listing at R.

It does not follow that all variables are so designated, and many famous named stars are variables. The brightest variable, for example, is the well-known Betelgeuse. Similarly many stars designated by Greek letters (with no ancient name still in use) are also variables, such as Delta Cephei (see Cepheids in Dictionary).

This system (or these systems) of Roman letters should not be confused with another usage of Roman letters for classifying stars according to their heat. The system was a development of the more general classification made by the nineteenth-century Italian astronomer Angelo Secchi, who divided the stars up into four types according to their spectra (essentially their basic colours): (1) white (or bluish), (2) yellow, (3) orange, and (4) red, the first of these being the hottest and the last (the red) the coolest. The Roman letter system to develop this was the one introduced in 1890 by the American astronomer Edward C. Pickering, director of the Harvard College Observatory. (Hence the system's original name of the 'Harvard system'.) He originally intended to use the Roman letters from A to Z to expand on the spectral and temperature classification, with A the hottest and bluest, and Z the coolest and reddest (and dimmest). But it was found unnecessary to have all alphabetical classes, so some were omitted. Moreover, the letter W was used for the remotest and hottest category, so that the run of Roman letters is now alphabetically disjointed and fragmented, and requires a mnemonic to memorise the sequence. The one normally used by astronomers is: 'Wow! Oh Be A Fine Girl Kiss Me Right Now, Sweetie' (or a final 'Smack', according to taste). This produces the sequence WOBAFGKMRNS. Stars in the 'W' category (the rarest) can have a surface temperature up to 80,000°C, while 'S' stars, which are merely reddish in colour, will be around 2600°C. Most stars are in the 'B' to 'M' range, with 'B' bluish-white or white in colour, 'F' and 'G' yellow, and the remaining three orange. But such designations either follow the name or occur separately in a sentence such as 'Sirius is an A star' or 'The Sun is a G star', so should not lead to confusion with the other usage already mentioned. (And the Harvard System, or Yerkes System, as it is now often called, is not used in this Dictionary anyway, but is simply mentioned here for interest and clarification.)

Apart from these names and designations, some stars still have popular names, almost folknames, such as the Dog Star and the Pole Star (for Sirius and Polaris), and although such names are treated in the Dictionary, they will not be used by astronomers. The more descriptive English names, however, such as Hind's Crimson Star (another variable) and Barnard's Star, where a particular astronomer is associated, will find their place in astronomical writing.

When it comes to the names of constellations, astronomers will almost always use the Latin name, and not talk of 'The Great Bear' but say 'Ursa Major'. This will apply even to the Southern Cross, which to astronomers is Crux Australis. Just six constellations, named after mythological characters, have names that are the same in both Latin and English: Andromeda, Cassiopeia, Cepheus, Hercules, Orion and Perseus.

Names of star groups and clusters, ranging from the Milky Way downwards, can be either English or Latin-based, in the latter case taking their name from the constellation with which they are associated, such as the Orion Nebula. Astronomers will use both, and what's more are perfectly happy with the somewhat trivial or unimaginative English names given to various nebulae, galaxies and other 'asterisms' (as star groups are sometimes generally called). So the Beehive Cluster, Clown Face Nebula, and The Fish Mouth, are just as acceptable as the most intricate star designation. (It will soon be apparent that such names normally indicate a fanciful resemblance to the particular object.)

However, nebulae are also designated in two other ways, in one or other of two catalogues. The Andromeda Nebula, for example, is also known as M 31 or NGC 224. The first of these means that it was assigned the serial number 31 in the catalogue compiled in the eighteenth century by the French astronomer Charles Messier; his designations are still in use for those nebulae that can be seen with small telescopes. The second designation has letters standing for 'New General Catalogue'. This was a more professional compilation of nebulae drawn up in 1888 by the Danish-born astronomer (working in Ireland) John Dreyer. It has been revised twice since his first listing, but is still very much in use, and in the southern hemisphere is essential, because Messier did not include any nebula south of declination −35°. Messier's original catalogue ended at M 103, but there are a few higher numbers added by later astronomers. NGC figures, meanwhile, are currently found well into four figures, and NGC 7662, for example, in Andromeda, is one of the brightest and easiest nebulae to see with a small telescope. Both Messier and NGC numbers feature where appropriate in this Dictionary.

Planet names are about the most familiar there are to non-astronomers, especially in the present age of space exploration (and, it must be said, exploitation), and astronomers use the same names as anyone else for them. Satellite names are similarly standard, although probably less familiar to the general public.

A point should be made here, however, about certain satellite names. Jupiter has (to date) sixteen satellites. In order of discovery they are (year of discovery in brackets): Io, Europa, Ganymede, Callisto (all 1610, and discovered by Galileo), Amalthea (1892), Himalia (1904), Elara (1905), Pasiphaë (1908), Sinope (1914), Lysithea and Carme (1938), Ananke (1951), Leda (1974), Adrastea (1979) and Thebe and Metis (1980).

These are mostly more or less obscure mythological names, of characters whose identity may or may not be immediately apparent. But there is more to it than that. The four satellites Ananke, Carme, Pasiphaë and Sinope are not only the outermost of the sixteen, but also the only ones to orbit in the opposite direct to their fellows, that is, they have retrograde motion. This peculiarity is indicated by the names themselves, as all four end in '-e', while the others (where of female characters) do not. Such usage of names for practical information is imaginative, without detracting from their traditional form.

The naming of Jupiter's satellites is an instructive history in itself. Before 1974, as can be seen from the above list, only twelve satellites were known, and of these, only the Galileans (the first four) and Amalthea had names, whereas the remaining seven were known in astronomical literature simply by the Roman numerals VI, VII, VIII, IX, X, XI and XII. This was better than nothing, of course, but it was clear they would soon need names.

Several suggestions for names were made. In 1962 the Soviet amateur astronomer E. I. Nesterovich proposed that satellite VI should be called Atlas, VII should be Heracles, VIII Proserpine, IX Cerberus, X Prometheus, XI Daedalus and XII Hephaestus. Meanwhile, names suggested by English-speaking astronomers were Hestia for VI, Hera for VII, Poseidon for VIII, Hades for IX, Demeter for X, Pan for XI and Adrastea for XII.

For many reasons neither list was regarded as satisfactory. For a start, many of the characters named have little connection in classical mythology with Jupiter (Zeus), which they obviously should. Moreover, Proserpine and Cerberus suggested more of a link with the planet Pluto, as 'underworld' names, while Atlas and Prometheus, as the names of Titans, presupposed some sort of connection with Uranus, their father. Even worse, Prometheus and Zeus (Jupiter) were mythological opponents, not allies! (Prometheus stole fire from Zeus, and was chained to a rock as a punishment, having his liver eaten by vultures daily, until he was rescued by Heracles.) Again, in mythology, Hades was an alternative name of Pluto, so falsely suggests an association with that planet, not with Jupiter.

So new names had to be found. In 1976, therefore, the Russian astronomer (and author: see Bibliography) Yu. A. Karpenko offered the following: VI as Adrastea, VII as Danaë, VIII as Helena, IX as Ida, X as Latona, XI as Leda and XII as Semele. The International Astronomical Union (IAU) responsible for the ratification of names meanwhile delayed its verdict. But then satellite XIII was discovered in 1974 and XIV in 1975, so whatever else was being proposed, it was clear that names would need to be established soon. The result was the names as we have them today, a fourth set of proposals, as can be seen. Satellite XIII, Leda, was named by its discoverer, the American astronomer Charles Kowal. All the others had names selected by the German philologist Johann Blunck. And this time, in one way or another, all the characters named did have a mythological link with Jupiter. (See the names of individual satellites in the Dictionary for details.) Of the most recent discoveries, Thebe's name was ratified by the IAU only in 1982, and the names of Adrastea and Metis were at first unofficial. (They were located on pictures taken from the Voyager space probe, and were designated respectively as 1979 J1 and 1979 J3, meaning that they were the first and third new satellites detected by Voyager 2 on its flyby of Jupiter in 1979. The probe was actually launched in 1977 and went on to fly past Uranus in 1986, where it recorded no fewer than *ten* new satellites to add to the five known, and was then programmed to continue on to Neptune. Their names, all Shakespearean in origin, are Cordelia, Ophelia, Bianca, Cressida, Desdemona, Juliet, Portia, Rosalind, Belinda and Puck.)

The story of the discovery of the minor planets (asteroids) has already been told, and details of their names will be found in full in Appendix II (p. 233) and in its Introduction. The names are little used by the general public, because the objects are themselves so small and local, but their names, together with their 'discovery number', are in regular use by astronomers, who will thus speak for example of '1578 Kirkwood' or '2602 Moore'. So many minor planets are constantly being discovered that it is impossible to name them immediately, or even designate them. But before a name is assigned to a particular asteroid, it will have a designation consisting of a combination of letters and figures. First will come the year of discovery, then a two-letter combination with possibly a subscript figure. The system is an ingenious one. Letters of the alphabet (Roman capital) are used to denote the particular half-month of the year in which the discovery was made, and the order of discovery in that half-month. (To denote the half-month, the alphabet is used without letters I and Z, but only letter I is omitted for the order.) If the order of discovery is greater than 25th (denoted by Z, the last letter of the alphabet), then the alphabet will have to be used again, and the subscript figure will show how many times it has been repeated. For example, asteroid number 1750, now named Eckert, was originally designated as 1950 NA_1, showing that it was discovered in the year 1950, in the first half of July (half-month 13, indicated by letter N, the 13th in this alphabet), and that it was the 26th to be discovered (as all 25 letters of the alphabet had been used up and the A here is the second time round). Readers might like to try their hand at a little asteroidal detection work here. The minor planet now known as 1484 Postrema was originally designated, before it was named in 1957, as 1938 HC. What can be deduced about its date and order of discovery? (Answer at the end of this Introduction!) See also *Hermes* for a special example.

Some minor planets have a high subscript figure, showing that many discoveries were made in a particular half-month. Asteroid 3269, for example, is designated as 1981 EX_{16}, showing that the alphabet had been gone through nearly 17 times to establish its order ($15 \times 25 = 375 + 23$ [letter X] = 398th in first half of March).

A few minor planets are designated simply with a four-figure number and the letter combination 'P-L', for example minor planet 3294 is 6563 P-L. This is merely a provisional designation, with the figure denoting its photograph number (2001 or higher, so as not to be confused with a year), and 'P-L' just meaning 'provisional'.

Far more spectacular than asteroids, however, and certainly far more visible to the naked eye, are comets. Who has not heard of Halley's Comet? Most comets are named after their discoverers such as, say, Donati's Comet, discovered by the Italian astronomer Giovanni Donati in 1858. Halley's Comet is not named after its discoverer, however, because it was well known, even notorious, many hundreds of years before his time, and it is named after him only because he predicted its return in seventy-six years (from when he observed it in 1682). This was the first time a comet's return had been predicted, and it is surely right that the comet should have been named after Halley, and not after its actual observer that particular year, G. Dorffel. On the whole, therefore, the Dictionary does not contain many comet names, unless they are exceptions to this rule, or are especially important for some reason.

Meteors, also visible to the naked eye, frequently appear in showers at particular times of the year, and they are usually named after the constellation from which they appear to radiate, with their shower named ending in '-id'. Thus the Ophiuchids radiate from Ophiuchus, the Taurids from Taurus, the Geminids from Gemini, and so on. (One shower, the Quadrantids, is named after an old constellation, no longer recognised, which is that of Quadrans Muralis.) There are over twenty known regular ('periodic') meteor showers, and most of them have their names included in the Dictionary, if only to confirm from which constellation they spring, and to give an indication of their time of appearance. (The Perseids are the most regular, and the Leonids the least. See their entries for the details.)

Meteor*ites* (which are not large meteors: see Glossary, p. 38) are named after the place on the Earth where they land, like the famous one that created the (misnamed) Meteor Crater in Arizona. Their names are not thus celestial but terrestrial, and as such they do not belong in this Dictionary. At least they get a mention here, however, as 'visitors from outer space'.

With such a vast, literally cosmic subject as this, the Dictionary will obviously have to be enormously selective in the names it includes and explains. Although most of the well-known and familiar ones are here, and many that are probably unfamiliar, there are literally thousands of names that could not be incorporated, and which would indeed occupy dozens of volumes on the subject, even if merely listed. The reader should know, though, that just about everything in space that can be named, because it has an identity of some kind, has been. This means that there exist, for example, maps of the surfaces of minor satellites that contain names more like those in a 'fantasy atlas' than of physical objects. But simply because the names are remote or recondite, that does not mean that they are random, and they will have been as carefully selected and reasoned as any others in the universe. On Jupiter's satellite Callisto, for example, the craters have names taken from Norse mythology, and include Adlinda, Alfr, Anarr, Asgard, Bran, Burr, Gloi, Grimr, Haki, Hodr, Igaluk, Lodurr, Loni, Nori, Nuada, Reginn, Rigr, Sudri, Tornarsuk, Tyn, Valfodr and (much more familiarly) Valhalla.

As previously mentioned, all celestial names have to be officially approved by the International Astronomical Union, a large organisation that includes thousands of the world's astronomers in its membership, and that holds a general assembly every three years to discuss new names (and many other astronomical topics). The

IAU is thus not just an astronomical organisation but an astronymical one, and its authority in the matter of celestial names is absolute. (It is quite capable of rejecting names, as well as accepting them.)

Most people will doubtless be unaware of this important role of the IAU, but may, however, have heard of certain current organisations that encourage members of the public to 'name a star' for a particular fee, and to issue the privileged namer a certificate to that effect. One of the best-known of these is the one called the International Star Registry, founded by a Canadian in 1979. He argued that as the stars did not belong to anyone, why could he not name the unnamed ones (of which there are thousands) and turn the venture into 'good business' at the same time? In practice what he does is to offer a designated star to a client, who then chooses a name for it (either his own, or selected from a list) and pays his fee. He then receives his certificate which will state, for example (to quote an actual instance): 'Know ye herewith that the International Star Registry doth hereby redesignate star number *Corona Borealis RA 16h 18m d 37°B* to the name *Her Royal Highness the Princess of Wales*. Know ye further that this star will henceforth be known by this name. Know ye further that this change is registered permanently'. (In this particular designation, 'RA' stands for 'Right Ascension', 'h' for 'hours', 'm' for 'minutes', 'd' for 'declination' and 'B' for the star's rating on the Yerkes scale, as detailed above. See the Glossary for other terms.)

Other 'redesignations' certified by the International Star Registry are much less dignified, and include the names of popular entertainers, private girl or boy friends, and 'pet' names on the lines of 'Hoppernicus', 'Boo Boo's Two Two' and 'Endless Entwining Ecstasy'.

Not surprisingly, the International Astronomical Union has taken rather a dim view of this activity, at one stage even trying to take the International Star Registry (whose very title cunningly reflects their own) to court, but in the event the case collapsed. Whatever the rights and wrongs of the matter, the fact remains that the International Star Registry does concern itself, however trivially or unprofessionally, with the naming (or renaming) of stars, so deserves a mention here. At least the issued certificate (available in 1986 for £25) does contain a true astronomical designation, and to that extent popularises star names and the world of astronyms. (The Registry, too, does not peddle popular astrology in the way in which many organisations and media personalities do.)

The spelling of the names of stars, and the rendering of their foreign language names, especially in Arabic, can be a tricky matter, and there can be a noticeable variation between the spellings of one source book and those preferred by another. For example, the star named Betelgeuse can also be spelt Betelgeux and Betelgeuze, and the original Arabic name of Alnilam can be rendered as Al Nithām or Al Naṭhm or (generically and more accurately) *an-niẓām*.

Which version should one take? The best principle here seems to be a rather delicate compromise between what is accurate and what is consistent. Therefore this Dictionary, on the whole, spells the current modern forms of the name as they appear in Ian Ridpath, *Guide to Stars and Planets*, and the Arabic names as they are given by Richard Allen in his *Star Names* (see Bibliography). The latter are not always accurate, and are often faulty in terms of modern Arabic, but at least they are consistent and are scholarly in their presentation. This Dictionary is a guide to the *origins* of astronomical names, and a fairly popular one at that, and not an academic examination of the various linguistic forms recorded for astronomical names over the centuries in different languages and dialects. (For an example of such a work, see the title by Kunitsch, who goes in some detail into variant name renderings and spellings and quotes the original Greek, Arabic, Hebrew, Latin, and so on. The book itself is in German.)

This Introduction is rather longer than some, because the different types of astronomical name, designation and nickname have had to be considered, and their disparity explained. I hope this will not deter readers from profitably using and enjoying the main body of the book, but on the contrary, may assist them in so doing.

Petersfield, Hampshire

Adrian Room

(Asteroid 1484 Postrema was the third to be discovered in the second half of April 1938.)

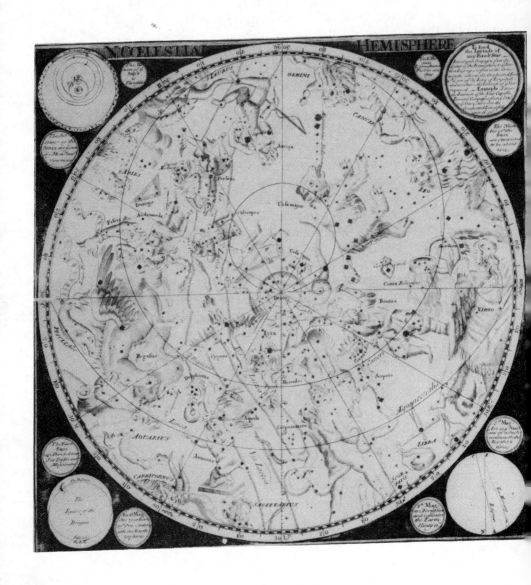

North Celestial Hemisphere, c. 1790, James Barlow.
(*Courtesy British Library*)

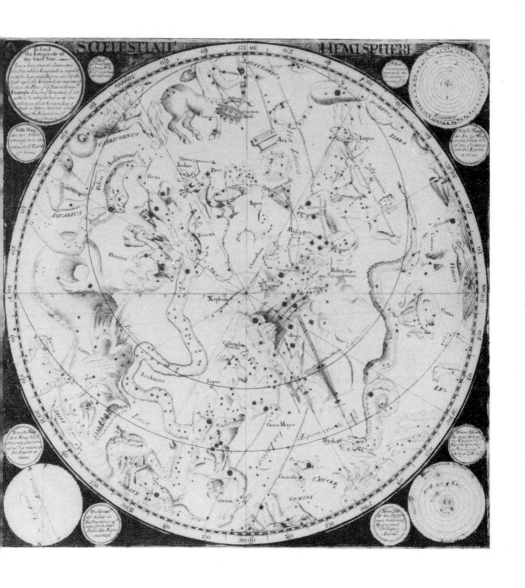

South Celestial Hemisphere, c. 1790, James Barlow.
(*Courtesy British Library*)

ASTRONOMICAL
GLOSSARY

The aim of this Glossary is to explain both the meanings of astronomical words and phrases occurring in the Dictionary and their origin. The latter can sometimes be a celestial eye-opener. Foreign terms, and terms with their own entries have been italicized.

asteroid
One of the thousands of small planets between Mars and Jupiter in the *Solar System*, also known as *minor planets*. (See Appendix II, p. 233, for a comprehensive listing of their names.) When they were first discovered, astronomers were uncertain what to call them, because they were both *planets*, yet were obviously not conventional planets. The Italian astronomer Piazzi at first suggested that they should be called 'planetoids' or 'cometoids', as their motion in the Solar System resembled that of planets or *comets*. But neither term caught on, even though some astronomers still use 'planetoid' today. It was the (German-born) English astronomer William Herschel who then proposed the more suitable term 'asteroids' or 'aoratoids', the latter based on Greek *aoratos*, 'invisible', as these bodies cannot be seen with the naked eye (with the sole exception of Vesta). He chose the name 'asteroids', he explained, from their 'asteroidical appearance', meaning that they were (literally) 'starlike'. His suggestion of 'aoratoids' then fell into disuse, and 'asteroid' won the day. It also is Greek based (*aster* means 'star'), with the '-oid' suffix meaning 'resembling' (itself likewise from Greek *eidos*, 'form', 'shape'). Although asteroids are not of course *stars* in the normal sense of the word, the term is reasonably apt. The Austrian astronomer Johann von Littrow had made another proposal – to call them 'zenareids'. This unusual name was a combination of the Greek names of Jupiter and Mars, that is, of Zeus (genitive Zenos) and Ares; the name thus indicated the position of the asteroids between the planets Jupiter and Mars. Not only was

his name over-contrived, but it was also too late, for the asteroids were already established under their present name.

astrology

As mentioned in the Introduction, 'astrology' formerly embraced the 'astronomy' of today and of the two terms, 'astrology' is the older, deriving from Greek *astrologia*, literally 'star-speaking'. (Compare other modern names of sciences, such as geology and biology, which are literally 'earth-speaking' and 'life-speaking', the 'speaking' now meaning simply 'study'.) It was thus only in the seventeenth century that 'astrology' became established in its modern, narrower sense of 'study of the stars with regard to their supposed influences on our daily lives'. 'Stars' here, too, really means either *constellations* or *planets*. Compare *astronomy*.

astronomical unit

The unit of length that equals the distance from the Earth to the Sun, which is roughly 149,600,000 kilometres (about 93 million miles). The rather clumsy term (usually abbreviated to 'a.u.') was devised in the early years of the twentieth century, and is used mostly for distances in the *Solar System*. Thus the average distance from Mars to the Sun is 1.524 a.u., and from Pluto 39.44 a.u.

astronomy

The branch of science that deals with the study of the *universe* and of the *stars*, *planets*, *comets* and much else featuring in this present Dictionary. The Greek word from which the name comes (*astronomia*) means literally 'star arrangement' (not 'star-naming', as is sometimes thought; compare *astronymy*). The word is a later one than *astrology* (which also see), and at first principally involved the mapping of the heavens. Subsequently 'astrology' and 'astronomy' were differentiated as effectively an 'art' on the one hand and a 'science' on the other, as the terms are used today.

astronymy

Not a common term, but one useful for this Dictionary, because it means 'the study of star names', or more generally 'the study of astronomical names'. (Compare 'toponymy' as a word meaning 'the study of place-names'.) A person who specialises in this is thus called an 'astronymist'. One needs to differentiate carefully here between 'astronymy' and *astronomy*. 'Astronymy' comes from the Greek words for 'star' (*astron*) and 'name' (*onyma*).

aurora

The 'polar lights', occurring when charged particles are emitted from the Sun. The respective names for the aurora at the North Pole and the South Pole are 'Aurora Borealis' and 'Aurora Australis' (these have their own entries in the Dictionary). 'Aurora' is a Latin word, meaning 'dawn' (or the name of the goddess of the dawn in classical mythology). The 'Aurora Borealis', or 'Northern Lights', seems to have been given its Latin title first in the seventeenth century by the French astronomer Pierre Gassendi. The name was suggested to him because the phenomenon resembled the sky at sunrise, when dawn is approaching and the horizon is illuminated. The Aurora Borealis was widely visible in France on 2 September 1621, when it was observed by Gassendi. The Aurora Australis was given its name subsequently, when a similar phenomenon was observed in the southern polar regions.

binary star

A system of two stars that revolve round each other. Such a system can also be called a *double star*, although this term additionally has another sense. 'Binary' means 'having two elements', and 'binary stars' seem to have been first described in the writings of William Herschel, in the late eighteenth or early nineteenth century. See also *eclipsing binary*.

celestial sphere

The 'celestial sphere' is the imaginary sphere that surrounds the Earth, with half of it formed by the apparent dome of the night sky, where all the astronomical bodies (*stars*, *planets*, and so on) seem to be projected. The centre of the celestial sphere is thus the same as that of the Earth's own globe. Anything described as 'celestial' in astronomy therefore relates to this imaginary sphere, or what is seen or observed or measured there. The 'celestial poles', for example, are the two points around which the stars seem to rotate daily, and the 'celestial equator' is the projection onto the celestial sphere of the Equator on the Earth, so that it is the great circle lying midway between the celestial poles. The 'celestial bodies' are all the stars, planets, *comets* and so on in the night sky (or even the day sky), and are what used to be referred to popularly as 'heavenly bodies'.

cluster

A 'cluster', or 'star cluster', is a group of faint stars that can be seen close together and that have properties in common, for example they share a common motion or are at a common distance from the Earth. Many clusters have *Messier* or NGC (*New General*

Catalogue) designations, not names, such as M 41 (NGC 2287), the large, bright cluster of about fifty stars in Canis Major. Of named clusters, the Pleiades is among the best known. Some clusters have more colloquial names, such as the Wild Duck Cluster (see both these in Dictionary).

comet

Comets can be defined as celestial bodies that follow an elliptical, even elongated *orbit* around the Sun in the *Solar System*, and that appear either periodically (at predictable intervals of just over 3 years to about 150) or non-periodically, that is, at such lengthy intervals that they cannot be predicted. When their orbit takes them near the Sun they have a characteristic 'tail' (pointing away from the Sun) and can sometimes be seen with the naked eye. They are sometimes popularly confused with 'shooting stars', and are regarded as racing across the night sky. (Their tail encourages this delusion, as does the name 'Comet' given to fast-moving vehicles such as jet aircraft. In fact they move only gradually against the background of the stars, as can be proved by observing one nightly.) The Ancient Greeks likened the 'tail' of a comet to a train of long hair, so they called it an *aster kometes*, literally 'hairy star', and it was from this latter word that modern English 'comet' developed (via Latin). Greek *kome*, 'hair', also lies behind the term 'coma' as the astronomical term for the head of the comet. (This is not the same 'coma' that means 'state of deep unconsciousness', which comes from Greek *koma*, 'deep sleep'.)

constellation

One of the *star* groupings seen from the Earth in an arbitrary configuration but usually having a special name applying to the depiction of a person, animal or other object that the configuration seems to represent. Many of the constellation names are very old, and among them are the famous *signs of the zodiac* beloved of astrologers. Today there are eighty-eight recognised constellations, with some of the best known in the northern celestial hemisphere. (In *astrology*, 'constellation' can also apply to the configuration of the 'stars', that is, *planets*, as they appear at the time of a person's birth.) The word came into English from Medieval Latin, formed from the prefix *con-*, 'with', and *stella*, 'star'.

declination

The celestial equivalent of terrestrial latitude, that is, the system of coordinates which, when used in conjunction with those of a body's *right ascension* (corresponding to longitude), determine its position in the *celestial sphere*. Declination is measured in degrees, minutes and seconds, and runs from the celestial equator (0°) in an arc to the celestial pole (90°). For an example of an actual declination reading, see *right ascension*.

double star

If two stars are linked together by gravity, so that they revolve round each other, they are specifically called a *binary star* (which see). If, however, they are not actually related but simply happen to lie by chance on the same line of sight, they are technically called an 'optical double'. Both types of system (each with two stars) are called a 'double star' (or simply a 'double'). To the naked eye, most stars appears as solitary objects, but in fact the majority have one or more 'companions', which can be seen only by telescope. The term double star was first used (in Greek) by Ptolemy, who described the star Eta Sagittarii as *diplous*. (One double star that can be detected with the naked eye is that of Mizar and Alcor: see these in Dictionary.)

dwarf

A 'dwarf' or 'dwarf star' is a type of star in the Main Sequence of stars that has a very high density and small diameter compared to a much brighter *giant*, which is not dense but diffuse. When this implied comparison is not made, a dwarf will be simply a 'standard' star (including the Sun) as distinguished from a *white dwarf*. Both dwarf and giant as opposing terms applied to stars were terms introduced by the Danish astronomer Ejnar Hertzsprung in 1905.

eclipsing binary

An 'eclipsing binary', or 'eclipsing *variable*', is a *binary star* with one of its components eclipsed (what is technically known as 'occulted') by the other, so that the total light we receive from the system is reduced. The best-known eclipsing binary is Algol (Beta Persei), where the eclipses occur just under every three days.

ephemeris

For astronomers, an 'ephemeris' is a special table that shows the predicted positions of a celestial body such as a *planet*, *asteroid* or *comet*. For an example of a well-known internationally recognised ephemeris, see the title in the Bibliography beginning '*Efemeridy. . . .*' The word itself is the Greek for 'diary' (itself comprising the prefix *epi-*, 'on', and *hemera*, 'day').

equinox
The word has two usages, one fairly general and the other more specifically astronomical. Generally the 'equinox' is one of the two occasions in the year, in spring and autumn, when the day and the night are of equal length. (Hence the word itself, which comes from the Medieval Latin for 'equal night'.) To an astronomer, however, an equinox is more likely to be the term for one of the two points in the *celestial sphere* where the path of the Sun (called the 'ecliptic') appears to cut the equator twice a year, in spring and autumn. The first of these, the vernal equinox, occurs on about 21 March, and at present lies in the *constellation* of Pisces, where it is also known (confusingly) as the 'First Point of Aries', because that was its position about 2000 years ago. (It is now in Pisces because of the effect of *precession*, and because of this same effect will eventually move on into Aquarius in about 2500 AD.) The autumnal equinox occurs on about 22 September and is now in Virgo. Its alternative name of the 'First Point of Libra', however, shows that formerly it was in that constellation.

galaxy
One needs to distinguish here between *a* galaxy and *the* Galaxy (with a capital 'G'). A galaxy is a vast collection of *star* systems held together by gravity, and usually spiral in form. It will thus contain not simply stars but *nebulae* and other interstellar matter. A well-known example is the Andromeda Galaxy, in the *constellation* of the same name. This is spiral-shaped, as is our own Galaxy, which is the galaxy to which our *Solar System* belongs. (Not all galaxies are necessarily spiral, however.) Our Galaxy is popularly known as the Milky Way, but this name can have another meaning and one needs to be careful when using it (see Dictionary). As explained in the Introduction (p. 6), the word 'galaxy' derives ultimately (via Latin) from Greek *gala*, genitive *galaktos*, 'milk'. Hence the link with 'Milky Way'.

giant
A 'giant' or 'giant star' is a type of very bright star with a very low density and a diameter up to one hundred times that of the Sun. Because of their colour, giant stars are often also called 'red giants', with one famous example being Arcturus. But not all giants are red, and Capella, for example, is yellow-white. The designation 'giant' contrasts this type of star with a *dwarf*. See also *supergiant*.

light year
A 'light year' is the distance that a beam of light travels in one year. As light travels at the speed of just under 300,000 kilometres a second, this means that one light year (l.y.) equals 9.46 million million kilometres, a truly astronomical figure that in itself may be so high as to seem meaningless. Most stars are thus several light years apart, and the nearest star to the Sun (Proxima Centauri) is 4.28 l.y. from it. (The Sun itself is about 30,000 l.y. from the centre of our *Galaxy*.) Compare *astronomical unit* for another way of measuring distances in space.

limb
The 'limb' is the edge of the apparent disc of the Sun, the Moon or a *planet*. It is not the same word in origin as the 'limb' that is an arm or leg, because it derives from Latin *limbus*, 'edge', whereas the bodily limb is an Old English word.

M number see **Messier number**

magnitude
A misleading term with a misleading scale! The 'magnitude' of a *star* does not relate to its size but is the measure of its brightness, and the lower the figure, the brighter the star. The system dates from 2000 years ago, but as it presently stands a star of magnitude 1 is defined as being exactly 100 times brighter than one of magnitude 6. (The Greek astonomer Hipparchus divided the stars that could be seen with the naked eye into six classes of brightness, with 1st magnitude being the brightest and 6th being the faintest visible. Our word 'magnitude' thus translates the original Greek *megethos*, which basically meant 'greatness'.) This means that stars brighter than magnitude 6 are given negative (minus) values, for example Sirius, the brightest star in the sky, has a magnitude of −1.46. By comparison, Regulus (Alpha Leonis) is of magnitude 1.35 as the 21st brightest star in the sky, while large telescopes have recorded stars with magnitudes down to 25. The magnitude of the Sun is −26.8.

mare (plural *maria*)
The word is Latin for 'sea' and is used for those areas of the Moon that were originally believed to be expanses of water. The term is still used although it has long been known that the Moon is waterless. The best description of a 'mare' in modern terms is as a dark lowland plain. The Latin word is found in many Moon names, and other Latin words also occur in names of other features that

resemble (or were once believed to be identical to) features on Earth. Among them are *oceanus*, 'ocean', *sinus*, 'bay', *palus*, 'marsh', and *lacus*, 'lake'. All these 'watery' designations are thus misnomers, and even the so-called 'craters' are really walled plains, and do not resemble the more familiar bowl-like depression in the top of a volcano or the deep 'bomb hole'.

Messier number
The system of identifying star *clusters*, *nebulae* and *galaxies* as they were recorded in the catalogue published in 1781 by the French astronomer Charles Messier. He numbered the Crab Nebula as 1, and then added further numbers for other nebulous objects up to 103, which is a cluster in Cassiopeia. Other astronomers later added further numbers up to 110. Messier's designations now appear in the form 'M 31' (for example), but all his numbered objects also have an 'NGC' number (see *New General Catalogue*), and this is the reference more generally used today. Many of Messier's objects additionally have modern nicknames, such as the Wild Duck Cluster (M 11), and they will be found in the Dictionary under these names.

meteor
A 'meteor' is a small particle of matter in the *Solar System* – believed to be the dust from a *comet* – that glows when it enters the atmosphere as the result of being heated by friction. It is then observable as the popular 'shooting star'. Meteors can appear either in isolation, as so-called 'sporadic' meteors, or in the form of a shower. In the latter case the shower usually is named after the constellation from which it seems to radiate (see Introduction, p. 25). The term itself comes from Greek *meteoron*, 'raised object', with the plural of this (*meteora*) understood to mean 'atmospheric phenomena'. The latter Greek word then gave, via Latin, the present 'meteor' or its equivalent in different languages. The 'atmospheric phenomena' sense can sttill be seen in the use of the term 'meteorology' to apply to the study of the weather. A 'meteorologist' is thus not a person who studies meteors in the same way that an entomologist, for example, studies insects, or an etymologist the derivations of words! Compare *meteorite*.

meteorite
A 'meteorite', as distinct from a *meteor*, is the word used for the remains of a celestial object, perhaps an *asteroid*, that has fallen to Earth through the upper atmosphere without being completely

burnt up. Meteorites are therefore not associated with *comets* in the way that meteors are. Their name is thus misleading, and is not helped by the fact that one of the most famous sites of a meteorite fall on Earth is called the 'Meteor Crater' (in Arizona, USA). The word is however formed directly from 'meteor', together with the '-ite' suffix found in names of minerals such as 'bauxite'. In the nineteenth century an alternative term arose for meteorites as 'aerolites'. This word was itself an alteration of 'aerolith', meaning literally 'air stone' (modern 'aero-' plus Greek *lithos*, 'stone'). Today 'aerolite' is reserved for those meteorites that are technically 'stones', as distinct from 'siderites' which contain iron, and 'siderolites' which are 'stony irons'.

minor planet
An alternative term preferred by many astronomers for an *asteroid*. However, some astronomers feel that the name implies that asteroids were formed from the debris of one of the *planets* proper, which is not so (they are the 'leftovers' of the planets in general when they came into being), and thus prefer 'asteroid' (which has a false suggestion of 'star'!). The term 'minor planet' seems to have arisen some time in the first half of the nineteenth century.

moon
An alternative word for a *satellite* of a *planet*, on an analogy with our own Moon. But most astronomers do not use the word in this meaning, if only to avoid confusion. (A few astronomers, though, claim that in modern times 'moon' meaning 'natural satellite' can make a distinction as against 'satellite' meaning 'artificial satellite'.) For the origin of the word, see 'Moon' in the Dictionary.

multiple star
A 'multiple star' is the equivalent of a *binary star* when there are more than two stars mutually attracted to one another by gravitation in the system. One classic example of a multiple star is Epsilon Lyrae, nicknamed the 'Double Double', which thus has four components.

nebula
Strictly speaking, a 'nebula' is one of the vast masses of gas and dust in space, such as those appearing in Messier's catalogue (see *Messier number*). Loosely, however, 'nebula' is used to mean *galaxy*, especially any galaxy outside our own (the Milky Way, see Dictionary). The term has been in use in English since the eighteenth century, and is taken from Latin *nebula*, 'mist' or 'fog',

itself related directly to Greek *nephele*, 'cloud'. Many nebulae have modern English nicknames, such as the Crab Nebula, the Dumbbell Nebula, and so on. (See these in the Dictionary.)

New General Catalogue

The New General Catalogue (NGC) is a listing, by number, of thousands of star *clusters*, *nebulae* and *galaxies* on the lines of Messier's catalogue (see *Messier number*) but far more comprehensive. It was compiled by the Danish-born astronomer John Dreyer in 1888 and is still in general use today, although now further backed up by two supplements called the Index Catalogues (IC). Many nebulous objects have both a Messier number and an NGC designation, with both given simultaneously. For example two galaxies in Ursa Major are referred to as 'M 81 (NGC 3031)' and 'M 82 (NGC 3034)'. Such double designations are used in this Dictionary for named nebulous objects. About 13,000 objects are now listed in the NGC together with its two supplements.

nova

A 'nova' is a faint *binary star* in which the *white dwarf* component explodes and suddenly increases in brightness, only to decrease to its original luminosity after a period of months or years. An example of a nova is the star designated HR Delphini that erupted (in Dephinus) in 1967. The word is Latin for 'new' (grammatically feminine as *stella*, 'star', is understood), and it was devised in the nineteenth century by Sir John Herschel for 'a star or nebula not previously recorded'. (Note that this is not its modern, more specialised sense, which came into use only in the twentieth century.) Compare *supernova*.

orbit

The curved path followed by one celestial body round another, such as the *planets* round the Sun or the Moon round the Earth. Most orbits are elliptical ('oval'). The word itself comes from Latin *orbis*, 'wheel' or 'circle'.

planet

One of the nine celestial bodies that revolve round the Sun in the *Solar System*, of which one is our own Earth. The word 'planet' is Greek in origin, for the Greeks referred to the planets as *asteres planetai*, 'wandering stars', as they appeared to 'wander' against the background of the fixed stars. Moreover, for the Greeks, the term applied not to the nine bodies we know today (three of which have been discovered only in comparatively recent times) but to seven other bodies. These were (in the order of their accepted distance from the Earth) the Moon, Mercury, Venus, the Sun, Mars, Jupiter and Saturn. We now list the planets (in the order of their known distance from the Sun) as Mercury, Venus, the Earth, Mars, Jupiter, Saturn, Uranus, Neptune and Pluto. (There are one or two favoured mnemonics for remembering this order. One for the space age is Moon Vehicles Easily Make Journeys Safe Under Nuclear Power.)

precession

As the Earth rotates in space, it very slowly wobbles, like a tilted spinning top. This means that the point in the sky to which the Earth's north and south poles are pointing is gradually changing, as is the position at which the apparent path of the Sun crosses the celestial equator (see *celestial sphere*). As the celestial equator is therefore slightly shifting all the time, so are the *equinoxes*. This gradual shifting (or the wobbling itself that causes it) is called 'precession', because each year the two equinoxes occur slightly earlier, thus 'preceding' the ones the year before. As a result of this, the vernal equinox, for example, moves by 50″ of arc every year, and has moved out of the *constellation* of Aries into Pisces. Overall, it takes about 26,000 years for the shift to return to its starting-point. One practical consequence of this, important for this Dictionary, is that the *signs of the zodiac* no longer coincide with the constellations after which they were originally named. Precession (the wobbling) itself is actually caused by the pull of the Sun and the Moon on the bulge round the equator of the Earth. The full term, 'precession of the equinoxes' (which alone makes astronomy worthwhile), is a translation of Copernicus's Latin *aequinoctiorum praecessio*. The Ancient Greeks called the phenomenon *metaptosis*, 'mutation'. (This should not be confused with the so-called 'nutation' or 'nodding' of the Earth's axis superimposed on its normal precession.)

pulsating star

A 'pulsating star', or 'pulsating variable', is a star that constantly changes its brightness in 'pulses' as a result of its expansion and subsequent contraction. (See *variable star* for a further reference.) A 'pulsating star' is not the same as a 'pulsar' (which 'pulsates' radio waves and is formed during a *supernova* outburst), even though this word evolved as a contraction of 'pulsating star'.

radiant
A radiant is the point in a *constellation* from which a *meteor* shower appears to radiate at regular intervals.

reflector
A reflector, or reflecting telescope, is a telescope that collects its necessary light by means of a mirror, as distinguished from a *refractor*. The first reflector was made by Newton in the seventeenth century.

refractor
A refractor, or refracting telescope, is a telescope which collects its light by means of a main lens (called the 'objective'), through which the light 'refracts' or 'bends' (as it does, for example, in a pair of binoculars). It is thus different from a *reflector*. Most amateur astronomers begin with a small refractor then move on, more seriously, to a reflector.

right ascension
The system of coordinates that is the equivalent of longitude on a terrestrial map and that is used, together with *declination*, to determine the position of a celestial object. Right ascension is measured in hours, minutes and seconds (usually abbreviated hrs, min, sec), as distinct from the degrees, minutes and seconds of declination. Here are the coordinates of Arcturus for the year 2000:

$$\text{RA } 14 \text{ hrs } 15 \text{ min } 39.6 \text{sec } \delta +19° \; 10' \; 57''$$

(The Greek letter, delta, is traditionally used for 'declination' and alpha is sometimes used instead of 'RA'.) The term 'right ascension' suggests that there is perhaps also a 'left ascension', but 'right' here means 'straight', as in 'right angle' (that is, not 'oblique ascension'). Declination values have a plus sign ($+$) for positions in the northern celestial hemisphere, and a minus sign ($-$) for positions in the southern hemisphere.

satellite
A relatively small celestial body that orbits round a (larger) *planet*, such as the Moon round the Earth. All the planets except Mercury and Venus have satellites, and all except the Moon have been discovered since the invention of the telescope. (Galileo led the way when he discovered four of Jupiter's satellites in 1610.) Satellites are still being discovered, and ten new ones were located for Uranus in 1986 (see p. 23). The word comes from Latin *satelles*, 'attendant',

and has its parallel in the Russian word *sputnik*, meaning 'fellow traveller' (but as used in English applying only to an artificial satellite, and in particular the first artificial satellite, launched by the Russians in 1957). The word *moon* is still sometimes used instead of 'satellite', as in 'Titan is Saturn's largest moon'.

selenography

The scientific term for the describing and mapping of the Moon, otherwise 'lunar geography', with the person who does this known as a 'selenographer'. The word is derived from Greek *selene*, 'moon'. The names of features on the Moon are sometimes called 'selenonyms' (compare *astronymy*).

sidereal period

The 'sidereal period' of a *planet* is simply the time it takes to revolve round the Sun, and of a *satellite* the time it takes to make one *orbit* of its primary planet. Thus the sidereal period of Mercury is just under 88 days (compared to the Earth's period of a fraction over 365 days), while Titan, Saturn's largest satellite, has a sidereal period of very nearly 16 days. The word comes from Latin *sidus*, 'star' or 'constellation' (see its use in the process of naming Uranus, in the entry for this planet in the Dictionary).

signs of the zodiac

The symbols that are conventionally used to represent each of the twelve zodiacal *constellations*, regarded as an essential adjunct to their art and craft by astrologers. The actual interpretation of the signs has been much discussed and disputed. If one takes the signs as the depictions associated with the various constellations, one can interpret them on a seasonal basis, with Aries and Taurus as spring signs, for example, Virgo as a summer sign, Libra and Sagittarius as autumn signs (weighing the fruits of autumn, for instance, or hunting game), and Aquarius a winter sign (representing rain). But the signs themselves are fairly recent, possibly of medieval origin only, and their interpretations are usually as given in the table which follows, where for convenience they are listed alphabetically, not in their traditional order. (The pictorial symbols sometimes used instead of the signs are also given.) See also *zodiac*.

Aquarius ♒

The undulating lines represent water. Aquarius is, after all, The Water Carrier.

Aries ♈
The symbol represents the head and horns of the animal that is The Ram.

Cancer ♋
The symbol appears to represent stylised claws or pincers of The Crab.

Capricorn ♑
The constellation's symbol is sometimes said to stand for the first two Greek letters (tau and rho) of the word *tragos*, meaning 'goat'. Others, however, prefer to see the representation of the Sea Goat's twisted tail, perhaps in fish-like form.

Gemini ♊
The symbol almost certainly represents two twins, standing together.

Leo ♌
The symbol is popularly regarded as representing the mane of The Lion. But could it not be more imaginatively seen as the animal's tail? It seems unlikely that the representation is of the initial Greek letter lambda, beginning *Leon*.

Libra ♎
What more obvious depiction than the beam of The Balance or The Scales?

Pisces ♓
Here the symbol represents the two Fishes, joined together.

Sagittarius ♐
A fairly obvious representation of an arrow, as shot by The Archer.

Scorpio ♏
The figure almost certainly represents the feet and tail (or sting) of The Scorpion. This symbol is very similar to that of *Virgo*, and there may have been some former link between the two, now lost. (Astronomers call this constellation Scorp*ius*.)

Taurus ♉
The symbol almost certainly shows the head and horns of The Bull.

Virgo ♍
A difficult symbol. It may represent a monogram of the first three Greek letters of *Parthenos*, 'virgin', although others see 'MV', for 'Virgin Mary' (!).

Solar System

Our immediate portion of the *Universe*, centring on the Sun and containing the nine *planets* together with their *satellites*, as well as the *asteroids* and any *comets*. All these bodies are attracted to the Sun by gravity, and any of them that shine (which all do) do so by reflecting the Sun's light. The basic unit of distance in the Solar System is the *astronomical unit*. 'Solar' is simply the adjective meaning 'of the Sun'. The term itself dates back to at least the beginning of the eighteenth century.

solstice

The Sun's path round the sky (called by astronomers the 'ecliptic') alters against the background of the stars in the *celestial sphere* as a result of the Earth's own *orbit* round the Sun. Moreover, the Sun's path is inclined at 23½° to the celestial equator, because the axis of the Earth is inclined at this same angle to the vertical. Therefore the Sun's path moves gradually north for one half of the year, and then gradually south for the other, to a maximum of 23½° in either direction. The most northerly point it reaches (around 22 June) is the 'summer solstice', and the most southerly is the 'winter solstice' (around 22 December). Hence the other sense of 'summer solstice' to mean the longest day of the year (the Sun is at its farthest point north, so can give the longest period of light) and of 'winter solstice' to mean the shortest day of the year (when the Sun is down in the southern hemisphere, so gives a minimum length of light). 'Solstice' itself is from Latin *solstitium*, literally 'sun stand-still', because when at these two extreme points north and south of the celestial equator, the Sun appears to 'stand still' and does not progress any further. Compare *equinox* (which is sometimes confused with it).

star

Today we differentiate between 'stars' and *planets*, reserving the former word for those celestial objects that appear as bright (or faint) twinkling bodies in the night sky, whether as individual, isolated objects, or in a diffuse mass. Formerly, however, star was used (as it still is, loosely or incorrectly) of any celestial body, whether a 'fixed star', as now, or a planet (such as Venus, the 'Morning Star' or 'Evening Star'), or a *meteor* (a 'shooting star'). This confusion is preserved and promoted by astrologers, who thus talk of a particular star presaging this or that event, when they are actually talking of a planet. In the broad astrological sense, too, 'the stars' means simply 'a horoscope'. The word itself, significantly enough, is an old one, and is similar in many languages, from Greek *aster* or *astron* and Latin *stella* to modern French *étoile*, German *Stern*, Welsh *seren* and so on. The original sense of the 'st-' (the 't' has dropped out in the Welsh word) was probably 'fixed', because the *st*ar is a body that *st*ays in a particular *st*ation in the sky. (These other English words have the same basic idea, as do words such as 'stand' and 'stop'.) Our nearest star is the Sun.

supergiant

A 'supergiant' is what it says: an outsize or 'super' *giant* star, that is extremely bright, and that has expanded to a diameter hundreds or thousands times greater than that of the Sun. A good example of a supergiant is Betelgeuse. The term arose only in the twentieth century.

supernova

A 'supernova' is not simply a large or 'super' *nova*. It is either a *binary star* in which the *white dwarf* gathers material from its companion and eventually destroys itself, or (more popularly) it is a *star* that has undergone a catastrophic (or cataclysmic) explosion as a result of using up its fuel, so that for a few days it becomes anything up to a hundred million times brighter than the Sun. One fine supernova to be seen in our own *Galaxy* was Kepler's Star of 1604, and the present Crab Nebula is the result of a supernova that was last seen on Earth in 1054. (It was recorded by Chinese astronomers, who observed it even in the daytime for just over three weeks, and at night for about two years.) In early 1987 astronomers were excited when a new supernova, the brightest since Kepler's Star, blazed out in the Large Magellanic Cloud.

Universe

For astronomers, the 'Universe' (usually with a capital 'U') is the sum total of everything in space, including all matter and energy and all celestial bodies. It is thus literally 'cosmic'. There have been many efforts to explain where the Universe came from and where it is going to. In brief, it is now generally believed that it arose in an instant 'big bang', began to expand, and is continuing to expand, presumably indefinitely. The word comes from Latin *universum*, in the sense 'the whole world', or literally 'turned into one', and it is also this word, and this sense, that lies behind the more down-to-earth 'university', which is thus a 'whole world' of education, complete with teachers and students, associated as a corporate body.

variable star

A 'variable star', or 'variable', is a *star* whose brilliancy changes over short periods. There are many different types. For one, see *eclipsing binary*. The term itself has been in use since at least the eighteenth century. See also 'Cepheids' in Dictionary.

white dwarf

Technically the *star* called a 'white dwarf' is not a *dwarf*, nor is it necessarily white. It is one of a type of small faint stars of very high density that has exhausted its supply of nuclear power and is thus in its final stage. One of the best known is the companion of Sirius, which it orbits once every fifty years. By coincidence, Procyon also has a white dwarf companion, so that this type of star is found accompanying the brightest stars in Canis Major and Canis Minor respectively. The term first came into use in the early twentieth century, and the difference between a white dwarf and a dwarf is emphasised by most astronomers, as in Jay M. Pasachoff and Marc Leslie Kutner's *University Astronomy* (W. B. Saunders, Eastbourne, 1978): 'Do not confuse the term "white dwarf" with the term "dwarf". The former refers to the dead hulks of stars, . . . while the latter refers to normal stars on the main sequence.'

zodiac

Astronomically the 'zodiac' is a belt, extending 8° to either side of the apparent path of the Sun (the 'ecliptic'), in which at any time the Sun, the Moon, the main *planets* and many of the best-known *constellations* are to be found, the latter being the twelve ancient constellations designated by the *signs of the zodiac*. Astrologically the zodiac is a circular diagram illustrating this belt, and showing the actual symbols representing these same constellations. As most of the depictions are of animals (see the Introduction, p. 3), the Greeks called this circle the *zoidiakos kyklos*, or 'circle of animal signs', ultimately from *zoion*, 'animal' (as in modern 'zoo'). Pluto is the only planet that can leave the belt of the zodiac. The English word zodiac has been in use since at least Chaucer's day. (By false association, the word is sometimes regarded as being related to 'zone', especially as the zodiac is a belt or zone. But the two words are quite distinct.)

DICTIONARY of ASTRONOMICAL NAMES

Acamar (star Theta Eridani)
The name is a corruption of *Achernar*, so has the same meaning, 'end of the river'. This is because originally *Eridanus*, as a 'river', wound its way down to this star, which was believed to be the 'end' point. In more recent times, Eridanus has been extended further south (to a point below the horizon, where it would have been invisible to the Ancient Greeks), and another star was given the name Achernar as the real 'end of the river'.

Achernar (star Alpha Eridani)
As mentioned above for *Acamar*, the title of Achernar was originally given to a star further up the 'river' of *Eridanus*. Subsequently this more southern star was assigned the name, which derives from Arabic Al Āhir al Nahr, 'the end of the river'.

Achilles (asteroid 588)
The asteroid was discovered in 1906 by the German astronomer Max Wolf as the first of the cluster known as the *Trojans*. He therefore gave it a name in honour of Achilles, the brave Greek hero of the Trojan War.

Acrux (star Alpha Crucis)
The name appears to be a compound of 'a' (for 'alpha') and '*Crux*' (the name of the constellation), and was possibly devised by the American astronomer Elijah H. Burritt for his Atlas published in various editions in the first half of the nineteenth century.

Acubens (star Alpha Cancri)
The name is a corruption of Arabic Al Zubanāh, meaning 'the claws', because the star is located at the end of the right-hand (southern) 'claw' of the crab that is the constellation *Cancer*. The star at the end of the other 'claw' is Beta Cancri and has no distinctive name.

Adhara (star Epsilon Canis Majoris)
The name means 'the virgins', from Arabic Al ʿAdhārā, with the allusion jointly to this star and others in *Canis Major*, especially perhaps *Wezen* and *Aludra* (with the latter name directly related to that of Adhara). The sense could be simply 'shining ones' or relate to some now obscure Arabic story.

Aeneas (asteroid 1172)
The asteroid is one of the *Trojans*, and so bears the name of one of the legendary heroes of the Trojan War. Aeneas was discovered in 1930 by the German astronomer Karl Wilhelm Reinmuth, who also discovered *Apollo* and *Hermes*.

Aestuum, Sinus ('bay' on Moon)
The name means 'Seething Bay' and is one of the old 'watery' names referring to the appearance of the Moon's surface in this area through an early telescope or even with the naked eye. In modern terms, the 'bay' is actually a marial area (see *mare* in the Glossary), lying to the west of the *Mare Vaporum*, with the 'land' formed by the *Apennines* (Montes Apenninus) to the north and the region extending southwards to the east of the 'bay'.

Agena (star Beta Centauri)
The star is located near the right foreleg of the centaur who lies behind the constellation of *Centaurus*, and appears to have a name based on rather dubious Latin meaning 'by the knee' (correctly *a genu*). An alternative name for the star is Ḥadar, Arabic for 'ground', perhaps for the star's low position.

Air Pump, The see **Antlia**

Albali (star Epsilon Aquarii)
The name derives from Arabic Al Saʿd al Bulaʿ, 'the good fortune of the swallower'. This strange name is said to have originally included the stars Mu Aquarii and Nu Aquarii, and to have indicated the fact that the two outside stars of this threesome (Epsilon and Nu) were farther apart than the stars Alpha and Beta Capricorni (respectively *Giedi* or *Algedi* and *Dabih*), so that they formed a wider 'swallower', in the sense that the outer two stars 'swallowed' the light of the third (that is, Mu Aquarii)! But this seems all too convoluted, and is probably an attempt to explain an obscure or distorted name.

Albireo (star Beta Cygni)
In the fanciful outline of a flying swan that is the constellation

Cygnus, Albireo is located at the bird's head, so might be supposed to have a meaning indicating this. However, the Arabian name for the star was different, as Al Minhar al Dajājah, although this does mean 'the hen's beak'. Perhaps, therefore, Albireo is a corruption of medieval Latin *ab ireo*, meaning 'from the rainbow', as suggested by some writers on star names. But others are doubtful.

Al Chiba (star Alpha Corvi)
The name is a version of Arabic Al Hibāʿ, 'the tent', which is one of the alternative Arabic designations for the constellation of *Corvus* as a whole, referring to its shape.

Alcor (star 80 Ursae Majoris)
This star, the close companion of *Mizar*, may have a name that comes from the same Arabic source as that of *Alioth* (Epsilon Ursae Majoris). If so, its name will thus be of the same uncertain origin. See *Alioth* for more detail.

Alcyone (star in *Pleiades* cluster)
The star is the brightest in the *Pleiades*, and has the name of the daughter of Atlas in Greek mythology who was one of the Pleiades (that is, her mother was Pleïone), and who was seduced by Poseidon. (This is not the same Alcyone as the one that married Ceÿx and that turned into a kingfisher, or 'halcyon'.) Among Arabic names for Alcyone were Al Jauz, 'the walnut', Al Wasaṭ, 'the central one', and Al Na'ir, 'the bright one'.

Aldebaran (star Alpha Tauri)
Partly for its orange colour, Aldebaran has been nicknamed 'Eye of the Bull', for its location as the left 'eye' in the head of the bull represented by the constellation *Taurus*. (It appears to be a member of the *Hyades* cluster, but is actually a separate star in the foreground.) Its name means 'the follower', from Arabic Al Dabaran, since it 'follows' the *Pleiades*, which are also in the constellation Taurus. Originally the name Aldebaran applied to the whole of the Hyades cluster, but has now been narrowed to this single star.

Alderamin (star Alpha Cephei)
The name is a corruption of Arabic Al Dhirāʿ al Yamīn, literally 'the right arm' (that is, of King Cepheus of Ethiopia, who is represented by the constellation of *Cepheus*). More precisely it appears as his 'right shoulder'.

Alexandra (asteroid 54)
The name looks a feminine one, of some woman called Alexandra.

In fact the asteroid, discovered in 1858, was named in honour of the German explorer and scientist, Alexander von Humboldt, who died the following year. See *mare Humboldtianum*.

Alfirk (star Beta Cephei)
The name is Arabic (Al Firk) for 'the flock', and may originally have applied to the whole constellation of *Cepheus*, because one star cannot represent a whole flock. For more on this pastoral picture, see *Er Rai*.

Algedi (star Alpha Capricorni)
The name, also rendered as Giedi, is a corruption of the Arabic title of the whole constellation of *Capricornus*, which was Al Jadi, 'the goat' or the 'ibex'.

Algenib (stars Alpha Persei and Gamma Pegasi)
The Arabic name means 'the side' (Al Janb), and is that of two stars in two different constellations, in each case referring to the extreme or 'side' location of each (in the case of *Pegasus*, the star is at the tip of the horse's wing). The star in *Perseus* has an alternative name of *Mirfak* (which see, for a more imaginative meaning).

Algieba (star Gamma Leonis)
The name is Arabic for 'the mane' (of a lion), Al Jabbah, and represents the position of the star at the point of the 'mane' on the crouching lion who is represented by the constellation of *Leo*.

Algol (star Beta Persei)
The star is famous as the first known eclipsing binary (see Glossary). Its English nickname of 'the Demon Star' is a fair rendering of the original Arabic, because Al Ghūl means 'the demon' (more precisely 'the ghoul', as a directly related word). In the constellation of *Perseus*, Algol represents the sinister winking eye of the Gorgon Medusa, who has been slain by Perseus and whose head is now held by him in his hand. The Arabs probably did not know of the 'winking', however, which was discovered by the Italian astronomer Montanari only in the seventeenth century.

Alhague (star Alpha Ophiuchi)
The name is also known as Ras Alhague, from the Arabic Rās al Ḥawwāʿ, 'the head of the serpent charmer'. This refers to the star's location at the 'head' of the man entwined by a serpent who is represented in the constellation of *Ophiuchus*.

Alhena (star Gamma Geminorum)
The Arabic name is Al Hanʿah, meaning 'the brand' or 'the mark',

especially one on the side of a camel or the neck of a horse. Here it appears to relate to a distinctive 'mark' in the feet of one of the twins, as they hold hands. See *Gemini* for the full picture.

Alioth (star Epsilon Ursae Majoris)
The meaning of the name is uncertain, although it appears to be a corruption of some Arabic word. Among the more improbable interpretations are 'the fat tail of the sheep' and 'the white of the eye', the latter said to have the sense 'very bright'. Perhaps it is not an Arabic word after all, so that the initial 'al-' is not the Arabic definite article ('the'). But if it is, the derivation could be in Arabic Al Jawn, 'the black horse', although the relevance of this is hard to explain. The name does appear, however, to have been borrowed by that of *Alcor*, also in the constellation of *Ursa Major*.

Alkaid (star Eta Ursae Majoris)
The star has the alternative name Benetnasch, and both names ultimately derive from Arabic Kā'id Banāt al Naʿash, literally 'governor of the daughters of the bier', that is, 'chief of the mourners'. This surprising name in the constellation of *Ursa Major* is due to the fact that the Arabs saw the outline of stars not as forming a bear but as resembling a coffin or bier. Alkaid (or Benetnasch) thus represents the chief mourner standing by the coffin. (The name Alkaid, incidentally, lies behind the English word 'alcalde', as a term for a chief official in a Spanish town, and in the name Benetnasch can be seen the root Arabic word which gave the derogatory English 'bint', as a slang term for a girl or woman.)

Alkalurops (star Mu Boötis)
The name is an Arabic rendering of Greek Kalaurops, meaning 'shepherd's staff' or 'crook'. The name of the constellation *Boötes* means 'herdsman' or 'cowherd', and the star represents a staff held in the herdsman's hand. With the Greeks, such a staff was used for throwing so as to drive back the cattle.

Almach (star Gamma Andromedae)
The star name, also spelt Alamak, possibly derives from Arabic Al ʿAnāk al ʿArḍ, literally 'the beast of the ground', referring to some predatory animal such as a badger. But this sense is hard to relate to the figure of *Andromeda*, and a more likely derivation is perhaps in Arabic Al Mauk, 'the sandal'. With this meaning, the star thus represents Andromeda's foot where it is chained to the rock to which she is bound.

Alnair (star Alpha Gruis)
The name derives from Arabic Al Nā'ir, 'the bright one'. This sense appears not to relate to the crane who is outlined in the present constellation of *Grus* but to the bright tail of the fish in the constellation of *Piscis Australis* when it earlier extended to take in Grus.

Al Nasl (star Gamma Sagittarii)
Sagittarius means 'the archer', and Al Nasl (more correctly Al Naṣl) is Arabic for 'the point'. The star is thus regarded as forming the point of the arrow held by the archer.

Al Nath see **El Nath**

Alnilam (star Epsilon Orionis)
The name represents Arabic Al Niṭhām (or Al Naṭhm), 'the string of pearls', and the star is seen as forming the basic 'jewellery' that decorates Orion's belt. (See *Orion* for the full story of the picture.) See also the names of the other two stars in his belt, *Alnitak* and *Mintaka*.

Alnitak (star Zeta Orionis)
The Arabic star name (correctly Al Niṭāk) means 'the girdle', and the star itself is the lowest of the three that form Orion's belt. See also *Alnilam* and *Mintaka*, as well as *Orion* itself for a fuller picture of the hunter.

Alpha Centauri
The designation (or catalogue description) is included here since it is one of the best-known classical star names (with a Greek letter), together with that of its companion, *Proxima Centauri*. For the origin of the star's Arabic name, see *Rigil Kentaurus*.

Alphard (star Alpha Hydrae)
The Arabic name Al Fard means 'the solitary one' (in full, Al Fard al Shujāʿ, 'the solitary one of the serpent'), and this denotes the prominent position of the star in a sparse region of the outline of the water snake (the constellation of *Hydra*), where it represents the creature's heart. It is the brightest star in the constellation.

Alphecca (star Alpha Coronae Borealis)
The name derives from Arabic Al Nā'ir al Fakkah, 'the bright one of the dish', and the star itself was seen as the central one in the crown that is represented by the constellation of *Corona Borealis*. Its alternative name of Gemma (a later Latin name not used in classical times) restates this in more modern terms.

Alpheratz (star Alpha Andromedae)
The star has the alternative name of Sirrah, and both these are derived from Arabic Al Surrat al Faras, 'the navel of the mare'. This refers not to *Andromeda* (!) but to the constellation of *Pegasus*, with which this particular star was formerly associated. In its present constellation, the star marks the head of Andromeda as she stands chained to the rock.

Al Rami (star Alpha Sagittarii)
The Arabic name (more exactly Al Rāmī) has the same meaning as that of the constellation of *Sagittarius* in which the star is located, 'the archer'. Its alternative (and more common) name is Rukbat, 'the knee', since this is what it represents on one of the legs of the centaur who is the archer. (The opposite knee is represented by two stars; see *Arkab*.)

Al Rischa (star Alpha Piscium)
The name, from Arabic Al Rishā', means 'the cord', with the star representing the knot that fastens together the tails of the two fishes who are seen in the constellation of *Pisces*.

Alshain (star Beta Aquilae)
The star name is probably an Arabic rendering of the Persian name for the whole constellation of *Aquila*, which was Shahin, literally 'royal', that is, 'royal bird' (the eagle that is Aquila) (compare English 'shah' for the King of Persia). The word also came to be applied to the falcon, especially the Indian species *Falcon peregrinator*. See also *Altair* and *Tarazed*.

Altair (star Alpha Aquilae)
The Arabic name is really one of those for the whole constellation of *Aquila*, in full Al Naṣr al Ṭāïr, 'the flying eagle' (literally 'the eagle the flying one'). Subsequently the name came to apply to each of the three stars Alpha, Beta and Gamma Aquilae: for the other two see *Alshain* (now known by its Persian equivalent name) and *Tarazed*. Readers of Lew Wallace's novel *Ben Hur* may perhaps recall that one of the horses in the famous chariot race belonging to Ben Hur's benefactor Sheik Ildarim was called Atair. This is basically the same name as that of the star, so means 'flying one'.

Altar, The see **Ara** (1)

Aludra (star Eta Canis Majoris)
The name derives from Arabic Al ʿAdhrā, 'the virgin', that is, the singular of Arabic Al ʿAdhārā, rendered in English as *Adhara* and

55

now used to designate the single star Epsilon Canis Majoris, although originally applying to a whole star group. In modern terms Aludra is a blue supergiant (see Glossary).

Alwaid (star Beta Draconis)
The name represents Arabic Al ʿAwāïd, meaning 'the mother camels'. This was originally a group name for the five stars Beta, Gamma, Mu, Nu and Xi Draconis, which formed the figure (later to be called in Latin Quinque Dromedarii). However, the name could also be taken as Arabic Al ʿAwwād, 'the lute-player', with the star group taken as showing such a musician. The star's alternative name today is Rastaban. This is usually explained as deriving from Arabic Al Rās al Thuʿbān, 'the head of the dragon', but has now been shown to derive from Al Rās al Tinnīn, with the same meaning, and with the same error as for *Thuban*. Compare *Eltanin*.

Alya (star Theta Serpentis)
The name appears to have the same origin as that of the Arabic name of the constellation of *Serpens* as a whole, although new scholarship now sees a derivation for the name in Arabic Al Alya, 'the sheep's tail'. Compare the name of the star Alpha Serpentis, *Unukalhai*.

Amalthea (satellite of *Jupiter*)
The fifth satellite of Jupiter was discovered by the American astronomer Edward Emerson Barnard (after whom *Barnard's Star* is named) in 1892. He himself did not give it any name, so to begin with the satellite was designated simply by the Roman number V ('five'). However, the French astronomer Camille Flammarion had proposed the name Amalthea for the satellite, and this was subsequently adopted. In Greek mythology, Amalthea was the goat who suckled the infant Zeus (Jupiter).

Amor (asteroid 1221)
Amor was discovered in 1932, and was given the name of the Roman god of love. This corresponds to the Greek name *Eros*, and doubtless the motivation was the same, because if the *Earth* is ignored, both Amor and Eros are between *Mars* and *Venus*. In classical mythology Eros was the son of Aries and Aphrodite, whose respective Roman equivalents were Mars and Venus. (Most of the asteroids, however, are between Mars and *Jupiter*.)

Ananke (satellite of *Jupiter*)
Ananke is the twelfth satellite of Jupiter, and like all the others so

far discovered has a name of classical origin. Most of the names do relate to the mythological Jupiter (Zeus), and Ananke was the name of the nurse who tended Zeus when he was still a baby. The satellite was discovered in 1951 by the American astronomer Seth Barnes Nicholson, who had also discovered *Sinope, Carme* and *Lysithea*. (See also *Pasiphaë* for the significance of the final '-e'.)

Andromeda (constellation and galaxy)
The constellation has an ancient name, that of the daughter of Cassiopeia in Greek mythology who was chained to a rock to be sacrificed to the sea monster Cetus as a punishment, when her mother claimed she was more beautiful than all the Nereids (daughter of Nereus) put together. However, she was saved by Perseus and subsequently married him. Certain named stars mark prominent points on the outline of Andromeda's figure, such as *Alpheratz* (her head), *Mirach* (her waist) and *Almach* (her foot). The Romans sometimes added descriptive words to her name when designating the constellation, for example Mulier Catenata ('the chained woman') or Virgo Devota ('the cursed virgin'). The Arabian astonomers called her Al Mar'ah al Musalsalah, with the same sense as the first of the Latin titles. Significantly, the constellations of *Cassiopeia* and *Perseus* lie respectively to the north and east of Andromeda in the sky. The Andromeda Galaxy, known officially as M 31 (in Charles Messier's Catalogue), lies within the constellation and was also known in ancient times.

Antares (star Alpha Scorpii)
The star is named not for its location in the constellation of *Scorpius* but for its colour. It shines with a conspicuous red glow, and so is called by a name which means 'rival to *Mars*' (the 'red planet'), from Greek Anti Ares, 'equal to Ares' (the Greek god of war who was the equivalent of Mars). The Arabic name for the splendid supergiant was Ķalb al ʿAķrab, 'heart of the scorpion', from the star's central position in the constellation.

Antennae (galaxies in *Corvus*)
The name is an unofficial one for the two galaxies officially known as NGC 4038 and 4039. They lie in the constellation of *Corvus* near the border with *Crater* and are so called because of their long antenna-like 'tails'.

Antlia (constellation)
Antlia, known in English as The Air Pump, is a somewhat obscure constellation in the southern hemisphere. It was so designated in

1763 by the French astronomer Nicolas Louis de Lacaille to mark the invention of the airpump by the English physicist Robert Boyle some hundred years previously. Lacaille's original name for the constellation was Machine Pneumatique, and this was at first rendered in Latin as Machina Pneumatica (echoing Boyle's own 'machina Boyleana', as he called it). Antlia itself is a name of Greek origin (in classical Greek, the word was used to mean both the hold of a ship and the bilge-water in it), with the term adopted in scientific Latin to mean 'pump'. Lacaille was responsible for several other constellation names of the 'sciences and fine arts' type.

Apennines (mountain range on Moon)
The range is the most conspicuous on the Moon, and was given its name, as were the Alps, by the seventeenth-century German astronomer Johannes Hevelius.

Aphrodite Terra (highland on *Venus*)
Aphrodite Terra is the largest highland area on Venus, approximating to the continent of Africa in size. Knowledge of the surface features of Venus is comparatively recent, and has been obtained chiefly only in the 1970s as the result of American and Soviet space probes. The names of features are therefore also modern. This particular highland bears the name of Aphrodite, the Greek goddess who was the equivalent of the Roman Venus, plus the Latin word for 'land'. Compare the name of the other prominent region, *Ishtar Terra*.

Apis see **Musca**

Apollo (asteroid 1862)
The very small asteroid was discovered in 1932 by the German astronomer Karl Wilhelm Reinmuth, who was to discover *Hermes* five years later, and it was given in a fairly arbitrary manner the name of the god of light (and much else) in Greek classical mythology. Apollo in turn gave its name to the particular family of asteroids here, whose orbit crosses that of the Earth, and which all thus have masculine names (see p. 235).

Apus (constellation)
The constellation, also known in English as The Bird of Paradise, was introduced in 1603 by the German astronomer Johann Bayer in his star atlas called 'Uranometria' (literally 'measurement of the heavens'). There is no obvious visual link between the outline of

the constellation and a Bird of Paradise, and it may simply be that Bayer gave the name purely on account of its exotic associations with the 'southern seas', because Apus is located near the south celestial pole. Bayer borrowed many of his new constellation names from the accounts of seafaring explorers, and he was clearly attracted to impressive bird names. Among others of his are *Grus*, *Pavo*, *Phoenix* and *Tucana*. (Confusingly, Bayer also appears to have introduced another constellation, now rejected, named Apis, 'The Bee'.)

Aquarids (meteor shower)
There are three annual meteor showers that radiate from *Aquarius*. The Eta Aquarids come in May, the Delta Aquarids in July, and the Iota Aquarids in August. The Greek letters denote the bright star that is closest to the radiant of each shower.

Aquarius (constellation)
The name is an ancient one, traditionally translated in English as 'The Water Carrier', and to the Babylonians the star outline represented the figure of a man pouring water out of a jar. The water jar itself is represented by the four stars *Sadachbia* (Gamma Aquarii) and the unnamed stars Eta, Zeta and Pi Aquarii. Aquarius is one of a number of 'watery' constellations in this area of the sky, and others nearby include *Capricornus*, *Cetus* and *Pisces*. It is thought that the particular association with water here developed because the Sun passes through the region during the rainy season, that is, late February to early March. The equivalent Greek name of Aquarius was Hydrochoös, literally 'water-pourer', while the Arabs called it Al Dalw, 'the well-bucket' and Al Sākib al Mā', 'the pourer of the water'. The so-called 'Age of Aquarius', foretold by astrologers, is still a long way off, and astronomically will not occur until about the year 2600 AD.

Aquila (constellation)
Aquila, or The Eagle, is an ancient constellation, with the bird itself represented by the outline of its stars. In Roman mythology, the eagle was Jupiter's constant companion, and (according to one account) carried his thunderbolts for him. The actual name of the constellation is echoed in the Arabic names of three of its stars, *Alshain*, *Altair* and *Tarazed*. Its corresponding Greek name was Aetos, also meaning 'eagle', as did the Arabic name of Al ʿOkāb.

Ara (1) (constellation)
The relatively little known constellation has a Latin name that

means The Altar. It has been known since classical times, and was regarded as representing the altar on which the Centaur (see *Centaurus*) was about to sacrifice the Wolf (*Lupus*). More generally, too, it was seen as the 'Altar of the gods'. Both Lupus and Centaurus lie to the east of Ara, although several of the stars here have subsequently been allocated to the constellation of *Norma*. The Arabic name for Ara was Al Mijmarah, 'the censer', suggesting that it was unknown to the Arabs in classical times.

Ara (2) (asteroid 849)
This name is nothing to do with that of the constellation (above)! The asteroid was discovered in the USSR and was given a name taken from the initials of the American Relief Administration, which granted relief to the USSR during the famine there of 1919–23.

Archer, The see **Sagittarius**

Arcturus (star Alpha Boötis)
The star is one of the brightest in the whole sky, and has a Greek name meaning 'bear-watcher' or 'bear-keeper'. This does not relate directly to the constellations of *Ursa Major* or *Ursa Minor* but to its own constellation of *Boötes*, the 'herdsman', who was seen not only as a driver of oxen but also as a hunter pursuing a bear (that is, the bear of Ursa Major). In fact, as with many other prominent stars, Arcturus shared an alternative name with its constellation, because it also appears in some texts and star charts as Boötes. The Arabic name for Arcturus was Al Simāk al Rāmiḥ, which has been translated as 'the lofty lance-bearer', so that the star itself must have been seen as the weapon in the hunter's hand. Today it is usually regarded as marking the edge of the hunter's tunic, so is some way near his feet.

Argo (former constellation)
Argo, or Argo Navis, usually known in English as The Ship Argo, was a former extensive constellation that was subdivided by the French astronomer Nicolas Louis de Lacaille in the 1750s into the four more manageable constellations of *Carina, Puppis, Pyxis* and *Vela* (in other words, the keel, poop, compass and sails of the original Argonauts' ship in Greek mythology). The Arabic name for the old constellation was Al Safīnah, 'the ship'.

Ariel (satellite of *Uranus*)
The smaller of the two inner satellites of Uranus, Ariel was dis-

covered in 1851 by the English amateur astronomer William Lassell. He named it after the chief of the sylphs 'whose humbler province is to tend the fair' in Pope's poem, *The Rape of the Lock*. The name happens to accord well with the 'fairy' theme.

Aries (constellation)

Known as The Ram in English, Aries is an ancient constellation, representing the ram in classical mythology whose golden fleece was sought by Jason and the Argonauts. The Greeks knew the constellation as Krios, with the same meaning, but the general Arabic name for it was Al Hamal, 'the sheep', and the Chinese called it by a name translating as 'the dog'. The Arabic name for the constellation is also seen directly in that of the star *Hamal* (Alpha Arietis). As represented in the sky, the ram is shown not with a butting head and horns, as might be imagined, but with his head turned to admire his own golden fleece (or to marvel at the bull in *Taurus* behind him).

Aristarchus (crater on Moon)

The crater is the brightest on the Moon, and it is named after the famous Greek astronomer Aristarchus of Samos, who was one of the first to suggest that the Earth revolved round the Sun and who tried to measure the relative distances of the Sun and Moon from the Earth.

Arkab (star Beta Sagittarii)

Arkab actually consists of two separate stars, known as Arkab Prior and Arkab Posterior, that can be distinguished by the naked eye. The basic name derives from Arabic Al 'Urkūb, 'the tendon' (that is, of the leg), so that the two names can be respectively understood as 'front part of the leg' and 'rear part of the leg'. *Sagittarius* is a centaur, with the body of a horse, and Arkab represents his left (or possibly right) forefoot, with *Al Rami* marking the other leg.

Arneb (star Alpha Leporis)

As frequently happens, one of the prominent stars of a constellation shares its name, and Arneb comes from Arabic Al Arnab, 'the hare', which is what the Latin name of *Lepus* means. Arneb lies in the centre of the crouching hare figure, where it is a yellow-white supergiant.

Arrakis (star Mu Draconis)

The name of Arrakis comes from Arabic Al Rāḳis, 'the dancer', probably with reference to a person (or animal, such as a camel) accompanying the star *Alwaid* (Beta Draconis), if this is interpreted as 'the lute-player' (see *Alwaid*). In the conventional representation of *Draco* as a dragon, however, Arrakis appears as its tongue (or possibly nose).

Arrow, The see **Sagitta**

Asellus Australis (star Delta Cancri)
Asellus Australis is one of a pair of stars, the other being *Asellus Borealis*. Their Latin names mean respectively 'southern donkey' and 'northern donkey', which at first sight appears to have nothing at all in common with the constellation of *Cancer* as a representation of a crab. The reference is probably to the nearby star cluster of *Praesepe*, 'the manger' (or 'the crib'), which the two stars flank. They can then be seen as two asses feeding at the manger.

Asellus Borealis (star Gamma Cancri)
This star is a twin to *Asellus Australis*, whose entry see for the probable interpretation.

Asterion (star Beta Canum Venaticorum)
Apart from *Cor Caroli* (Alpha Canum Venaticorum), Asterion is the only star of any prominence in the constellation of *Canes Venatici*, the 'hunting dogs', and its name means simply 'starry', from the fainter stars which surround it. Its southern companion is Chara, 'dear one', 'delight' (both Latin *cara* and Greek *chara* can be seen here). The two names together are taken to be those of two dogs held on a leash by *Boötes* as they pursue the Great Bear (*Ursa Major*) round the pole. See all these names for further details about the hunter and the bear, and see also *Taurus*!

Astraea (asteroid 5)
Astraea was discovered in 1845 by the German amateur astronomer Karl Ludwig Hencke, who two years later also discovered *Hebe*. The first asteroids to be discovered were given the names of goddesses in classical mythology, and Astraea was assigned that of the goddess of justice.

Auriga (constellation)
The constellation, known since Babylonian and Greek times, has a name usually translated in English as The Charioteer. No particular charioteer seems to be portrayed, although he has been regarded by various astronomers in the past as representing Erichthonius, son of Hephaestus (Vulcan). who inherited his father's lameness and so needed a chariot as locomotive. Auriga also had several

Greek titles, all related to chariot-driving, such as Heriochos ('rein-holder'), Armelates ('charioteer') and Helasippos ('horse-driving'). The traditional outline of stars does not actually include a chariot, however, but simply its driver, a young man holding a whip. The most prominent star of the constellation is *Capella*, the 'she-goat'. See also *The Kids*.

Aurora Australis ('Southern Lights')
The Latin name is used for the atmospheric phenomenon, observed around the South Pole, in which curtains or streamers of coloured light are seen to move across the sky. (The spectacular glows are produced by charged particles from the Sun, which penetrate the outer air in the upper atmosphere of the Earth.) The corresponding phenomenon around the North Pole is the *Aurora Borealis*. Aurora was the name of the Roman goddess of the dawn, and the atmospheric displays resemble a colourful 'dawning'. Australis and Borealis correspond respectively to the classical names of the south wind and north wind, Auster and Boreas. (By coincidence, the flow of charged particles from the surface of the Sun is now known as 'solar wind'.)

Aurora Borealis ('Northern Lights')
The atmospheric phenomenon probably more familiar to readers of this book than the *Aurora Australis*, its counterpart in the southern hemisphere.

Australe, Mare ('sea' on Moon)
The name means 'southern sea' (see, however, *mare* in the Glossary) and the region so designated is virtually the only one of its type in the mainly upland region of the south-east section of the nearside of the Moon, where it is near the limb (see Glossary).

Baily's Beads (phenomenon on Moon)
Baily's Beads is the name given to the brilliant points or 'beads' of sunlight seen briefly around the edge of the disk of the Moon just before and after a total eclipse. They are caused by the Sun shining through valleys in mountainous regions on the limb (see Glossary). The phenomenon is named after the English amateur astronomer Francis Baily, who observed it during the total eclipses of 1836 and 1842. (However, the occurrence had been remarked some time before this and was described at least a century earlier, for example, by Edmond Halley, of comet fame.)

Balance, The see **Libra**

Barnard's Star (in *Ophiuchus*)
The star, technically a red dwarf (see Glossary) known as Munich 15040, is named after the American astronomer Edward E. Barnard, who identified its rapid motion in 1916. Barnard's Star has the largest known proper motion (relative to the other stars), and is believed to have planets of its own. Barnard had earlier discovered *Amalthea*.

Beehive Cluster see **Praesepe**

Bellatrix (star Gamma Orionis)
The star's Latin name means 'female warrior', and is itself probably a rendering of the Arabic name Al Najid, 'the conqueror'. This in turn may well have originated as an alternative name for *Orion* himself.

Benetnasch see **Alkaid**

Berenice's Hair see **Coma Berenices**

Betelgeuse (star Alpha Orionis)
The famous red supergiant star has a name of early Arabic origin that indicates its location in the outline of *Orion* as he stands brandishing his club and shield. This is Yad al Jauzāh, literally 'hand of the giant' (Orion). Many current astronomical works, however, give the meaning as 'armpit of the giant'. This is an error caused by European translators working from Arabic into Latin, when they misread Arabic *bad* (modern *bāṭ'*, 'armpit') for *yad* (modern *yadd*, 'hand'), as the letters vary by only a single dot. Hence the incorrect 'B-' as the first letter of Betelgeuse, and the name today should really be something like 'Yedelgeuse' (compare the names of *Yed Posterior* and *Yed Prior*). Although lettered Alpha, Betelgeuse is not as bright as Beta Orionis (see *Rigel*).

Bettina (asteroid 250)
The asteroid, discovered in 1886, bears the name of Baroness Rothschild, who had subsidised its discovery and who wished to have her first name commemorated astronomically.

Biela's Comet
A former comet which used to return every 6.75 years but which broke in two on its return in 1845 and is now regarded as defunct. It is named after the Austrian astronomer who first identified it in 1826, Wilhelm von Biela.

Big Dipper, The see **Ursa Major**

Bird of Paradise, The see **Apus**

Black Eye, The (galaxy in *Coma Berenices*)
The famous spiral galaxy, officially known as M 64 (NGC 4826), is so named because its nucleus is edged by a dark cloud of dust, giving a fanciful resemblance to a black eye.

Blaze Star, The (in *Corona Borealis*)
The star is officially known as T Coronae Borealis. It is so named because although it usually 'slumbers' as a recurrent nova (see Glossary), it can suddenly and dramatically brighten from about magnitude 11 to magnitude 2. This last happened in 1946, and its next 'blaze' cannot be accurately predicted. (Its previous one was in 1866.)

Blinking Planetary Nebula, The (in *Cygnus*)
Officially known as NGC 6826, the nebula has been given its name simply because it appears to blink on and off. The 'blinking' effect can be seen by looking alternately at the 10th magnitude star at its centre and away from it. A fairly powerful telescope will be needed for this observation, however.

Blue Planetary Nebula, The (in *Centaurus*)
The small nebula, with official number NGC 3918, was so named by the astronomer Sir John Herschel, who discovered it. The nebula is blue in colour and resembles *Uranus* when seen through a telescope, although larger than this planet.

Boötes (constellation)
Today the name of the constellation is usually given in English as The Herdsman, because the stars represent a herdsman driving a bear round the sky. The bear is *Ursa Major*, and the herdsman is often regarded as holding the leash of two hunting dogs, the *Canes Venatici*. Moreover, the name of *Arcturus*, the constellation's brightest star, also means 'herdsman'. But at the same time the name can be interpreted as 'ox-driver' or 'cowherd', from the two Greek words that mean respectively 'ox' and 'drive'. In some older English texts, therefore, the constellation is called The Wagoner. Again, others have regarded the name as deriving from Greek Boetes, meaning 'clamorous', referring to the shouts made by the driver to his oxen. But if Boötes is seen more as a huntsman, these shouts will have been to urge on his hounds! This 'shouting' concept is seen in some Latin versions of the name, such as Vociferator or Clamator. Alternative Greek names have been found as Arcto-

phylas, 'bear-keeper', and Arctouros, 'bear-guard' (compare *Arcturus*). In Arabic, the constellation has been called *Nekkar*, which is regarded as a corruption of Al Baḳḳār, 'the herdsman'. Overall, the reference to a bear makes more sense than to an ox, because there is no neighbouring constellation with a bovine name, but Ursa Major *is* adjacent.

Border Sea, The see **Marginis, Mare**

Brucia (asteroid 323)
The asteroid was discovered in 1891 by the German astronomer Max Wolf, and was named after the American benefactress Catherine Bruce, financial backer of several important astronomical projects. Brucia was one of the first asteroids to be named after a woman.

Bull, The see **Taurus**

Caelum (constellation)
The small, almost inconsequential constellation bordering on *Eridanus* was introduced in the 1750s by the French astronomer Nicolas Louis de Lacaille, with a name that is usually translated as The Chisel (but also known as The Burin or The Graving-Tool in some former English texts). Its alternative Latin name was Caela Sculptoris, or Scalptorium, which suggests a false association with the constellation of *Sculptor* (also introduced by Lacaille).

Callisto (satellite of *Jupiter*)
Callisto is the outermost of the four main satellites of Jupiter first studied in detail by Galileo in 1610 (see *Galileans*), and it was named, as were the other three, by the German astronomer Simon Marius. In Greek mythology Callisto was the nymph changed into a she-bear by Artemis for not remaining a virgin, like her other attendants.

Caloris Basin (on *Mercury*)
The Caloris Basin is the largest structural feature on Mercury and consists of a vast crater containing several smaller (and more recent) craters. It lies near a 'hot pole', or the region with the highest temperature on the planet, hence its name (with Caloris being Latin for 'of heat').

Camelopardalis (constellation)
The constellation, whose name is also rendered as Camelopardus, is a decidedly faint and rather obscure one. In English it is known

as The Giraffe, and its long and straggling outline can perhaps be imagined to have some resemblance to this animal. The Latin name was apparently devised for it by the Dutch astronomer Petrus Plancius some time in the early seventeenth century, but it was not prominent until recorded in a book written in 1624 by the German mathematician Jakob Bartsch, a son-in-law of the famous astronomer Johannes Kepler. It contains no named stars.

Cancer (constellation)
The name is an ancient one, and Cancer (called The Crab in English) represents the particular crab that attacked Hercules in Greek mythology when he was wrestling with the *Hydra*. The wretched crab was crushed underfoot by Hercules but was subsequently 'promoted' to the heavens, where it remains in the form of the present constellation. (This all occurred as part of the Second Labour of Hercules.) As a name, Cancer is the Latin word for 'crab', but the Greeks also knew the constellation by the same designation, calling it Carcinos. The Arabs, too, called it by the same name, as Al Saraṭān, 'the crab'. The 'claws' of the crab are represented by the two stars *Acubens* (Alpha Cancri) and the unnamed Beta Cancri. In olden times, the Sun reached its most northerly point in the sky when it was in Cancer, achieving this on 21 June, which came to be called the 'summer solstice' (the latter word meaning literally 'sun station'). At this time, the Sun appears overhead at noon at the latitude (on Earth) of 23.5° north. This latitude came to be called the Tropic of Cancer, a name still in use today, despite the fact that because of the effect of precession (see Glossary) the Sun is actually in the constellation of *Gemini* at the time of the summer solstice. Because of the undesirable association of the name Cancer, some popular astrologers (as distinct from astronomers) prefer to call the constellation by its English name. See also *Capricornus* for more on the tropic.

Canes Venatici (constellation)
The constellation, whose Latin name means The Hunting Dogs, was introduced in 1690 by the German astronomer Johannes Hevelius. The dogs are represented by the star called *Asterion* or *Chara*, and are regarded as being held on a leash by *Boötes* (as hunter rather than herdsman) as they chase their quarry (*Ursa Major*, the Great Bear) round the pole. Originally Hevelius had given the two names to two separate stars, the other being the one now known as *Cor Caroli* (Alpha Canum Venaticorum).

Appropriately Canes Venatici is flanked in the night sky by Boötes on one side and Ursa Major on the other.

Canis Major (constellation)

Known as The Greater Dog in English, Canis Major is an ancient constellation representing, together with *Canis Minor*, the dogs who followed the heels of *Orion* in classical mythology. The Latin name was borrowed by the Romans from the Greeks, who called it either Cyon, 'the dog', or Astrocyon, 'dog star', with both these names referring not to the constellation as a whole, however, but to its brightest star, *Sirius*. Hence also the further Greek name of Seirios. The Greeks did not thus distinguish between a 'greater' dog and a 'lesser'. Arabian astronomers called the constellation similarly, Al Kalb al Akbar, 'the greater dog' (literally 'the dog the greater'). In the sky, Canis Major lies next to *Orion*, as one would expect, with *Lepus* also crouched at his feet.

Canis Minor (constellation)

Canis Minor, known in English as The Lesser Dog, represents the second of the two dogs of *Orion*, the other being *Canis Major*. Just as the Greeks called the larger constellation by the name of its brightest star, *Sirius*, so they knew Canis Minor as *Procyon*, literally 'before the dog', because it 'rose' in the sky before its larger equivalent. (See *Procyon* for more about the single star Alpha Canis Minoris.) The Arabs referred to the constellation not only as Al Kalb al Asghar, 'the lesser dog', translating the Latin title, but also as Al Shi'rā al Shāmiyyah, 'the bright star of Syria'. The latter name refers to the fact that the constellation (or its brightest star Procyon) disappeared from view when it set over Syria. The two constellations of Canis Major and Canis Minor are not quite contiguous in the sky, because they are divided by the more recently introduced constellation of *Monoceros*.

Canopus (star Alpha Carinae)

The star has a name direct from classical mythology, because Canopus was the pilot of the fleet of the Greek king Menelaus, and the constellation in which it is located is *Carina*, The Keel (but originally *Argo Navis*). (The piloting reference is coincidentally apt today, as Canopus is used as a guide for navigating spacecraft.) Arab astronomers, however, knew the star as Suhail, usually understood as deriving from Al Sahl, 'the plain', this being a personal title indicating someone who was handsome or admirable in some way. (As a Muslim personal name today, Suhayl is seen as meaning 'easy', 'smooth'.)

Capella (star Alpha Aurigae)
The brightest star in *Auriga* has a Latin name meaning 'little goat', and the Greeks identified Capella with Amalthea, the she-goat in classical mythology who suckled the baby Zeus (see *Amalthea*, the satellite of *Jupiter*, for more on this). An early Arabic name for the star was Al Rakib, 'the rider', because it appeared in the evening sky before the other stars and so seemed to 'drive' them up. In the outline of the charioteer, who is Auriga, Capella marks the driver's left shoulder, while his right shoulder is marked by Beta Aurigae, *Menkalinan*.

Caph (star Beta Cassiopeiae)
The star's name derives from the Arabic name of the constellation of *Cassiopeia* as a whole, which was originally Kaff al Ḥadib, 'hand (stained) with henna'. Caph itself actually marks the upper right-hand corner of the chair on which Cassiopeia sits (see *Cassiopeia*).

Capricornus (constellation)
Capricornus, often called in English The Sea Goat, as well as Capricorn, is an ancient constellation whose outline was seen as depicting a goat with a fish tail. However, the Latin name itself literally means 'goat horn', that is, 'horned goat', and was itself a translation of the Greek name, Aigoceros or Aigocereus. The Arabic name was Al Jadi, 'the goat', and this is the basis for the name of the constellation's most prominent star, known as *Algedi* or *Giedi* (Alpha Capricorni). (See also its Delta star, *Deneb Algiedi*.) Over two thousand years ago, the Sun used to reach its most southerly point below the (earthly) equator in Capricornus on 22 December, which was called the 'winter solstice'. The latitude of 23.5° south, at which the Sun appears at this solstice, was therefore known as the Tropic of Capricorn, and we have kept the name even though, because of precession (see Glossary), the winter solstice has now moved into the adjacent constellation of *Sagittarius*. Compare *Cancer* for an analogous situation in the northern hemisphere. Capricornus, with its fish tail, adjoins a number of constellations that also have 'watery' associations, such as *Aquarius, Cetus* and *Pisces*.

Carina (constellation)
Originally Carina was part of the large constellation of *Argo Navis*, the Ship of the Argonauts. In the mid-eighteenth century, however,

the latter was subdivided into four new constellations by the French astronomer and cartographer Nicolas Louis de Lacaille. These were Carina (The Keel), *Puppis*, *Pyxis* and *Vela*. For an appropriately named star in Carina, see *Canopus*.

Carme (satellite of *Jupiter*)
Carme was discovered in 1938 by the American astonomer Seth Barnes Nicholson, who also discovered *Lysithea* (the same year), *Ananke* and *Sinope*. In Greek mythology, Carme was the mother by Zeus (Jupiter) of the Cretan martyr Britomartis. (See also *Pasiphaë* for the significance of the final '-e'.)

Cassini's Division (in the rings of *Saturn*)
The name is that of the main division in the rings, between the two bright outer rings, lettered A and B, and the less prominent inner ring, lettered C. (Further rings have subsequently been discovered by space probes in the 1970s.) The Division is named after the Italian astronomer who discovered it in the seventeenth century (as well as four of Saturn's satellites), Giovanni Domenico Cassini.

Cassiopeia (constellation)
In Greek mythology, Cassiopeia was the beautiful but boastful wife of *Cepheus*, King of Ethiopia, and mother of *Andromeda*. Because of her constant boasting she was elevated to the heavens in the form of her present constellation, where she is depicted seated in a chair and holding up both her arms in supplication. The constellation is easily recognised by the five bright stars that form the 'W' of her chair. (This is valid for the northern hemisphere: in the southern hemisphere the stars form an 'M'.) The Greeks and the Romans both had the same name for Cassiopeia, although the Greeks also sometimes referred to the constellation as He tou thronou, 'she of the chair'. This latter name is similar to the one given by the Arabs, which was Al Dhāt al Kursiyy, 'the lady in the chair', although they also called the constellation Kaff al Hadib, 'the hand (stained) with henna'. The second of these names thus relates to quite a different figure: a hand with five fingers, of which two are henna coloured, especially the unnamed Eta Cassiopeiae, which is a double star with yellow and red components. See also *Caph* for a related name, that of the constellation's Beta star, as well as *Kaffaljidhmah* (Gamma Ceti) for a similar 'hand' concept.

Castor (star Alpha Geminorum)
Gemini means The Twins, that is, Castor and *Pollux*, the twin sons of Leda, and the two young men are duly represented not only in

the outline of the constellation as a whole, but also in the names of its two most prominent stars, respectively Alpha and Beta Geminorum. The twins have their practical uses, too, for the stars are known to be 4.5° apart and so can be used for determining angular distances. This is appropriate for the two men who were members of the Argonauts' crew in Greek mythology.

Cat, The see **Felis**

Centaur, The see **Centaurus**

Centaurus (constellation)
The outline of the constellation depicts a centaur in Greek mythology, that is, one of the creatures who had the head, arms and upper body of a man but the lower body and legs of a horse. Many Greeks believed that the constellation represented Chiron, the wise centaur who was the tutor of many Greek heroes, and who was raised to the heavens after being accidentally hit by a poisoned arrow from Hercules. To the Arabs, some of the constellation's stars showed a different outline, named by them as Al Ḳaḍb al Karm, 'the branch of the vine'. However, they also recognised the depiction of the centaur, and called the main constellation Al Kentaurus. Centaurus has three named stars: *Rigil Kentaurus* (Alpha Centauri), *Agena* or *Hadar* (Beta Centauri), and the companion of *Alpha Centauri* that was discovered only in the twentieth century, *Proxima Centauri*. This last name is the most widely used of the three today.

Cepheids (type of variable star)
Most stars have a constant brightness. Some alternate, however, from being bright to being dim, with the period of variation ranging from an hour or so to several years. There are different types of 'variables', as they are called, but the Cepheids are properly short-period variables, or pulsating stars, with a range of variation from just a few days to a few weeks. There are variables of this type in different constellations, but the Cepheids are named after their prototype, which is Delta Cephei, in *Cepheus*.

Cepheus (constellation)
Cepheus is an ancient constellation, named after the mythological King Cepheus of Ethiopia, husband of *Cassiopeia* and father of *Andromeda*, whose respective constellations adjoin it. Cepheus is represented in his royal robes, with one foot on the pole and the other on the so-called 'solstitial colure', the great circle on the

celestial sphere that passes through the solstice (see Glossary). The king's head is represented by the three stars Delta, Epsilon and Zeta Cephei, which form a triangle. The name of the constellation was the same for the Greeks, Romans and Arabs, with variations in spelling depending on the language. Cepheus is famous for its pulsating variable star Delta Cephei, which gave the *Cepheids* their name. For an Arabic view of the constellation, see also *Alfirk* and *Er Rai*.

Ceres (asteroid 1)

Ceres was the first asteroid to be discovered, the pioneer discoverer being the Italian astronomer Giuseppe Piazzi, who came across the asteroid in 1801 when compiling a star catalogue. Ceres, which is the largest asteroid, but not the brightest (which is *Vesta*), is named after the Roman goddess of fertility (Greek Demeter). The asteroid was not named immediately, and the French astronomer Joseph Lalande at first proposed that it should be called 'Piazzi', after its discoverer. (Napoleon wanted to call it 'Juno'!) Piazzi himself, however, proposed the name Cerere di Ferdinando, 'Ferdinand's Ceres', after the reigning King Ferdinand of Naples and Sicily. But the name was shortened by scientists to simply 'Ceres'. The particular goddess was chosen because Ceres was regarded as the agricultural patroness of Sicily, and because Sicily was where Piazzi was born.

Cetus (constellation)

The English name of the constellation is The Whale, and in classical mythology Cetus was the name of the sea monster who threatened to devour Andromeda before she was rescued by Perseus. The outline of Cetus in the night sky does actually look something like a 'monster of the deep', and can be seen suitably basking on the bank of its adjacent constellation, the river *Eridanus*. The Greeks called the constellation Orphis or Orphos, as well as Cetos, with this being the standard word for a type of large fish (usually translated into English as 'sea-perch'). The Romans borrowed the Greek name direct, as did the Arabs, who knew the monster as Al Ketus. Some named stars in the constellation refer to parts of its body, such as *Menkar* (Alpha Ceti, its nose) and *Deneb Kaitos* (Beta Ceti, its tail). See also *Kaffaljidhmah*, the name of Gamma Ceti.

Chamaeleon (constellation)

The name, which means what it says, is not an ancient one, as are many names of creatures and animals, but was introduced for the rather unremarkable constellation in the southern skies by the German astonomer Johann Bayer in 1601. None of its stars are named, at least in western countries, and the actual outline of the formation only vaguely suggests a chameleon. (It is really more like an elongated coffin.)

Chameleon, The see **Chamaeleon**

Chara see **Asterion**

Charioteer, The see **Auriga**

Charles's Oak see **Robur Carolinum**

Charles's Wain see **Ursa Major**

Charon (satellite of *Pluto*)
Charon is the one known satellite of Pluto, discovered by the American astronomer James Christy only in 1978. The name is apt, because in Greek mythology Charon was the ferryman who transported souls across the river Styx to Hades, also known as Pluto, the god of the nether world. (Hades' name was actually transferred to this kingdom of the dead.) There is more to the name than might appear, however, because in choosing it for its mythological appropriateness, Christy also had the name of his wife Charlene in mind! (Exactly the same sort of link between the name of the celestial object and that of its namer can also be found in Pluto itself, so doubtless Christy aimed to mirror this, too.) Charon should not be confused with *Chiron*.

Chiron (asteroid 2060)
Chiron may not actually be a true asteroid, although at present it is classified as such. It was discovered in 1977 (only a year before *Charon*) by the American astronomer Charles Kowal. The object puzzled him, because until then hardly any asteroid had been found with an orbit beyond that of *Jupiter*. But whether true asteroid or not, it was named after the learned centaur Chiron in Greek mythology (see *Centaurus*), perhaps with an additional hint at Charles Kowal's own name.

Circinus (constellation)
Circinus is one of the relatively obscure constellations introduced in the southern sky by the French astronomer Nicolas Louis de Lacaille in the mid-eighteenth century. Its Latin name means The Compasses, which it does up to a point resemble, with its pair

73

of 'legs', and Lacaille undoubtedly brought it in to accompany neighbouring *Norma* (The Level).

Clock, The see **Horologium**

Clouds, Sea of see **Nubium, Mare**

Clown Face Nebula (in *Gemini*)
The name (or nickname) refers to the appearance of the nebula when viewed through a fairly large telescope. In a small telescope it appears simply as a greenish disk. The nebula is also called the Eskimo, for the same reason: its pale 'face' with contrasting darker 'eyes' and 'mouth'. Properly it is NGC 2392.

Coalsack Nebula (in *Crux*)
The nebula of this name in fact overflows into the neighbouring constellations of *Centaurus* and *Musca*. The nickname, which has also been Soot Bag in English, refers to the dark appearance of the nebula as compared to the relative brightness of the surrounding stars. It has had various other names in different languages over the years, such as Latin Macula Magellani ('Magellan's spot'), Italian Canopo fosco ('dark net') and English Inky Spot. (An Australian version saw it as an emu, and an embodiment of evil.)

Coathanger, The (group of stars in *Vulpecula*)
The small group comprises six stars almost in a straight line, with a 'curve' of stars (the 'hook' of the hanger) extending from its centre. The Coathanger can be easily observed through binoculars, and will be found near the border of the constellation with that of *Sagitta*.

Cold, Sea of see **Frigoris, Mare**

Columba (constellation)
The constellation's Latin name translates as The Dove in English, and was probably envisaged originally as representing the dove that followed Noah's Ark, as the full name was usually given as Columba Noachi. However, such a Christian (or at least biblical) reference is unusual in constellation names, so that many astronomers have preferred to look for a classical allusion, and have found it in the dove that the Argonauts sent ahead to help guide them safely through the so-called Symplegades, the 'clashing rocks' that were supposed to close in on all ships that sailed through them. This second interpretation makes additional sense for the location of Columba adjoining *Puppis* (which was after all originally part of

Argo Navis). The name of Columba is not an old one, however, and was first formally recognised in 1679, when it appeared on a list compiled by the French astronomer Augustin Royer. It had been included in some earlier star charts, though, in particular those made by the German astronomer Johann Bayer.

Coma Berenices (constellation)
A more romantic constellation name than most, representing Berenice's Hair, as it is usually rendered in English. The outline of the stars was thought to suggest the flowing tresses of Queen Berenice of Egypt, who cut off a lock of her hair as a 'thanksgiving' to the gods when her husband Ptolemy Euergetes returned safely from battle (after invading Syria to avenge the death of his sister, also called Berenice). The lock of hair then mysteriously disappeared, but turned up in the sky as the present constellation. The story is an old one, and based on fact (Berenice was put to death by her son Ptolemy Philopator in 221 BC), but the constellation was not established by the name until 1602, when it was included in a star catalogue compiled by the Danish astronomer Tycho Brahe.

Compass Box, The see **Pyxis**

Compasses, The see **Circinus**

Cone Nebula (in *Monoceros*)
The nebula, which is within the range of good amateur telescopes, is so called because of its tapering shape. It lies in the combination of star cluster and nebula officially known as NGC 2264.

Cor Caroli (star Alpha Canum Venaticorum)
The Latin name means 'heart of Charles', and was given to the star in 1725 by Edmond Halley (of the comet), as Astronomer Royal, in honour of Charles I, who had been executed in 1649. The star was supposed (but without foundation) to have shone particularly brightly in 1660, when Charles I's son, Charles II, succeeded to the throne and the monarchy was restored. The unusual royal name fits somewhat awkwardly into its classical and mythological setting, and can of course mark no feature related to the two dogs of the *Canes Venatici*. The Arabic name for the star was Al Kabd al Asad, 'the liver of the lion'. This is not a literal sense but a symbolic one used for the highest star in any constellation or grouping as reckoned from the equator.

Corona Australis (constellation)
The constellation, so named from ancient times, was regarded as a

counterpart to the *Corona Borealis*, so that it is The Southern Crown. Precisely which crown is intended, if any, is uncertain, but one legend associated the name with the crown worn by the centaur who is the main figure of *Sagittarius*, and Corona Australis lies next to this constellation. Ptolemy's Greek name for it was Stephanos Notios, 'the southern wreath'. The later Arabic name for the constellation had the same meaning as the Latin (and English), and was Al Iklīl al Janūbiyyah. Earlier Arabic names varied, however, and included Al Ḳubbah, 'the tortoise', and Al Ḣibāʿ, 'the tent', both referring to its outline. Unlike Corona Borealis, it contains no named stars.

Corona Borealis (constellation)
The ancient constellation has a name translating into English as The Northern Crown. Originally it was simply The Crown (in the relevant language), because *Corona Australis* was named subsequently in relation to it. Thus its Greek name was simply Stephanos, 'the wreath', with Boreios ('northern') or Protos ('first') added to this in due course to distinguish the two. The Arabs also came to call it by a name corresponding to the present Latin and English one, and this was Al Iklīl al Shamāliyyah. However, an earlier Arabic name had been Al Fakkah, 'the dish', with reference to the outline of the constellation, and it is a corruption of this that today still gives the name of the star Alpha Coronae Borealis, which is *Alphecca*. It is still uncertain which crown the ancient astronomers and astrologers had in mind. Perhaps it was the one that Bacchus gave to Ariadne as a wedding present, and that he threw into the heavens when she died. Or possibly it was the one that Theseus gave Ariadne after she had helped him enter the labyrinth and kill the Minotaur. Whichever it represented, it can be clearly seen today as a 'golden curve' of seven stars, with *Gemma* (alias Alphecca) set in the centre as a real jewel.

Corvus (constellation)
Corvus is Latin for The Crow, and this is now the accepted English name. The constellation has an ancient name, but the Greeks knew it slightly differently as Corax, 'the raven', and the old Arabic name, Al Ḳurāb, meant the same. In classical mythology, the crow was sent by Apollo to collect water in a cup. Instead it spent its time eating figs, later returning to Apollo holding a water snake in its claws and claiming that this creature had been the cause of its delay. Apollo, angry at the deception, banished all three, crow, cup and water snake, to the heavens, where they are now found as

the respective constellations of Corvus, *Crater* and *Hydra*, grouped together. The name of the star Alpha Corvi, *Al Chiba*, relates to the whole outline.

Crab, The see **Cancer**

Crab Nebula (in *Taurus*)
The name may seem to suggest that the nebula is somehow connected with the constellation of *Cancer* (The Crab). In fact it is nothing to do with it. It was positively identified in the eighteenth century by more than one astronomer, but was not named until 1848, when the astronomer Lord Rosse (William Parsons), observing it through his 72-inch (183 cm) telescope, thought it resembled a crab's claw. The Crab Nebula is actually the remains of a star that exploded as a supernova (see Glossary).

Crane, The see **Grus**

Crater (constellation)
The constellation is an ancient one, and its equally Greek and Roman name means The Cup. In mythology, it represents Apollo's goblet (see *Corvus* for the story). Its corresponding Arabic name was Al Baṭīyah (which is more exactly an earthern jar for wine). The Latin word that gave the name is that used for the 'craters' on the Moon, because they are seen as bowl-like.

Crêpe Ring (of *Saturn*)
The nickname is used for the third of the three main rings of Saturn that is officially known as Ring C. It is so called because it is semi-transparent, like crêpe paper. It is also known as the Dusky Ring, because it is not as bright as the two other chief rings, A and B.

Crises, Sea of see **Crisium, Mare**

Crisium, Mare ('sea' on Moon)
The Latin names translates into English as 'Sea of Crises', and it first appeared in the map of the Moon published in 1651 by the Italian astronomer Giovanni Riccioli, who however based his material on the work of his pupil, Francesco Grimaldi. It was one of a number of similar names relating to human emotions and experiences, and is to be understood more as 'Sea of turning-points' than 'Sea of decisions'. Doubtless, too, it was intended to be contrasted with some other name of opposite sense, such as the *Mare Tranquillitatis*, or one of the 'land' names (with Terra) that have not survived. Certainly the *mare* itself, which is in the east

central section of the near side of the Moon, is not literally a 'sea of crises', because it has remarkably few craters.

Crommelin's Comet
A comet with a return period of just over twenty-seven years, and named after the English astronomer (of French descent) Andrew Claude de la Cherois Crommelin (died 1939), who correctly deduced that a comet observed previously by four different astronomers was one and the same. It last appeared in 1984.

Crow, The see **Corvus**

Crux (constellation)
The name is Latin for 'cross', but the usual English name for the constellation is The Southern Cross, sometimes represented by the fuller Latin title of Crux Australis. The name denotes its location in the southern hemisphere, and there is no official 'northern cross' with which it can be contrasted, although 'Northern Cross' is a popular nickname for *Cygnus*. It was formed out of *Centaurus* in the sixteenth century by a number of astronomers and navigators, so is not an ancient name. Although the smallest constellation in the sky, it is also one of the most distinctive, with its four stars Alpha (*Acrux*), Beta, Gamma and Delta Crucis forming the four arms of the cross. These stars were visible to Greek navigators in the Mediterranean in ancient times, but the effects of precession (see Glossary) mean that the constellation has now been carried below the equator so that it cannot be seen in Europe.

Cup, The see **Crater**

Cursa (star Beta Eridani)
The name looks Latin, but is actually Arabic in origin, from Al Kursiyy al Jauzah, literally 'the chair of the central one'. This refers to the position of the star under the foot of *Orion*, which lies next to *Eridanus* in the sky.

Cygnus (constellation)
The Latin name is translated as The Swan, and the constellation was seen in ancient times as a swan flying down the *Milky Way*. It is hard to say which particular swan, if any, was intended in the original legends, for there are several classical tales involving the bird. Perhaps it was the swan into which Zeus turned when he visited Leda, wife of King Tyndareus of Sparta, with the result of their union being the heavenly twin *Pollux*. The Greeks knew the constellation as both Cycnos, 'the swan', and simply Ornis, 'the

bird', but the Arabs usually called it Al Dajājah, 'the hen' (see *Albireo* in this connection, the name of the star Beta Cygni). Many of the star names in Cygnus refer to the different parts of the bird, such as *Deneb* (its tail), *Gienah* (its wing) and *Sadr* (its breast). See also *Crux* and *Cygnus Loop*.

Cygnus Loop (in *Cygnus*)
The Cygnus Loop is a nebula in Cygnus, between the star Epsilon Cygni and the border of the constellation with *Vulpecula*. The second word of the name refers to the swirling shape of the gaseous nebula, which is the remains of a former supernova (see Glossary). The brightest part of the Loop is called the *Veil Nebula*.

Cynosura see **Polaris**

Dabih (star Beta Capricorni)
The Arabic name in its full original form was Al Sa'd al Dhābiḥ, literally 'the lucky one of the slaughterers'. The reference is to the ritual sacrifice of animals by heathen Arabs at the rising of *Capricorn* (when the animal would often be a goat, although not exclusively). The name of Dabih was also applied to the star Alpha Capricorni (see *Algedi*).

Deimos (satellite of *Mars*)
Deimos is the smaller of the two satellites of Mars, the other being *Phobos*. Both satellites were discovered in 1877 by the American astronomer Asaph Hall, who named them after the sons of Ares (Mars) in classical mythology who were constant companions of their father and who often drove his chariot into battle. Their Greek names are suitably warlike (literally martial), and mean respectively 'terror' and 'fear'.

Delphinus (constellation)
The corresponding English name of the constellation is The Dolphin. The name itself goes back to classical times, thus incidentally showing man's long association with this particular marine animal. In mythology, dolphins served as messengers of Poseidon, god of the sea, and in one story saved the life of his son Arion, when he was attacked on board a ship. The Greek name of the constellation, with this meaning, was Delphis or Delphin, although the Greeks also knew it as Hieros Ichthus, 'the sacred fish'. The later Arab name was Dulfīm, also meaning 'dolphin', although originally the Arabs called the constellation by other names, such as Al Ḳa'ūd, 'the camel' (for riding), or Al 'Uḳūd, 'the pearls', this

latter name applying in particular to the four stars Alpha, Beta, Gamma and Delta Delphini, the first two of which are individually named as *Sualocin* and *Rotanev*. These same four stars have also been seen as forming a distinctive shape other than that of a dolphin, hence their alternative name or nickname as Job's Coffin.

Deneb (star Alpha Cygni)
The Arabic name means simply 'tail', because this star represents the tail of the swan who is seen in the constellation of *Cygnus*. The fuller Arabic name for the star was slightly different, however, and was Al Dhanab al Dajājah, 'the tail of the hen'. Compare the other 'tail' (Deneb) names.

Deneb Algiedi (star Delta Capricorni)
The Arabic name means 'tail of the goat', and describes the location of the star at the end of the fish tail of the goat who is depicted in the constellation of *Capricornus*. The proper Arabic form of the name was Al Dhanab al Jady: compare the name of Alpha Capricorni, *Algedi*, as well as the other Deneb names.

Deneb Kaitos (star Beta Ceti)
The Arabic name means 'tail of the whale', and refers to the position of the star at this point on the outline of a whale that can be seen (by some) in the constellation of *Cetus*. The full Arabic name was originally Al Dhanab al Ḳaiṭos al Janūbīyy, that is, 'the tail of the whale to the south'. This referred to its position by comparison with another Deneb Kaitos, the star now known only as Iota Ceti, which was Al Dhanab al Ḳaiṭos al Shamāliyy, 'the tail of the whale to the north'. Today many astronomers refer to the stars as Diphda, a corruption of its alternative Arab name of Al Ḍifdiʿ al Thānī, 'the second frog', with *Fomalhaut* being alternatively Al Ḍifdiʿ al Awwal, 'the first frog'. Compare the other Deneb names.

Denebola (star Beta Leonis)
Like the other Deneb names, this star represents the tail of the creature depicted in the constellation. In this case the Arabic origin is Al Dhanab al Asad, 'the tail of the lion', the latter being *Leo*.

Dew, Bay of see **Roris, Sinus**

Diadem (star Alpha Comae Berenicis)
The star is the only named one in the faint constellation *Coma Berenices*, or Berenice's Hair, and its name indicates its status as the main 'jewel' at the base of the queen's flowing locks.

Dione (satellite of *Saturn*)
Dione is the fourth of Saturn's many satellites, and was discovered in 1684 by the Italian astronomer Giovanni Cassini, who had already discovered *Iapetus* and *Rhea* some years before. Many of the satellite names of this planet are those of the Titans in classical mythology, with these being the first-born children of *Uranus* and Ge and the brothers and sisters of Cronus, that is, of Saturn. Dione was the mother of Aphrodite, the daughter of Zeus. See also *Titan* himself, as the name of the largest of Saturn's satellites, and the first to be discovered.

Diphda see **Deneb Kaitos**

Dipper, The see **Ursa Major**

Dog Star, The see **Sirius**

Dolphin, The see **Delphinus**

Dorado (constellation)
The constellation was introduced in 1603 by the German astronomer Johann Bayer. Its name, unusually, is Spanish, literally meaning 'golden', 'excellent', but also being the word for a spiny-finned fish of the genus *Coryphaena*, whose scales change colours but include glints of gold. The English name of the constellation, on the other hand, is The Goldfish, or equally The Swordfish. The latter of these reflects its Greek name of Xiphias, used by Halley and others in the seventeenth century. It has no named stars, although it does include the *Tarantula Nebula* and its encompassing Large *Magellanic Cloud*. Its outline in the sky does perhaps suggest a scintillating fish of some kind.

Doris (asteroid 48)
The asteroid was discovered in 1857 by the German astronomer Hermann Goldschmidt, who also discovered *Nysa* and *Europa*. It is named after Doris, the mother of the Nereids in classical mythology. Some of the names of her fifty daughters occur as the names of other asteroids, including ones discovered in earlier years.

Double Double, The (star Epsilon Lyrae)
This is the nickname, rather than official name, of a rare type of star, a quadruple, that is, two double stars (see Glossary). The star is not only the finest of its kind in the constellation of *Lyra* (it is the only one there), but the finest of all quadruples in the night sky.

Dove, The see **Columba**

Draco (constellation)
The name is Latin for The Dragon, and it is an ancient one, denoting the creature whose body lies coiled round the north (celestial) pole. Dragons feature in many mythologies. This classical one is possibly the one slain by Hercules (called Ladon in the story of his Eleventh Labour), when it guarded the garden of golden apples of the Hesperides. At any rate, as depicted in the sky, *Hercules*, in the neighbouring constellation, kneels with one of his feet on the dragon's head, that is, on the stars *Alwaid* (Beta Draconis) and *Eltanin* (Gamma Draconis). The Greek name for the constellation was the same, as Dracon, and this was translated by the Arabs as Al Tinnīn and Al Thuʿbān (see the name of *Thuban*, the star Alpha Draconis).

Draconids see **Giacobinids**

Dragon, The see **Draco**

Dreams, Lake of see **Somniorum, Lacus**

Dschubba (star Delta Scorpii)
The name represents Arabic Al Jabhah, 'the forehead', because it is located at this point on the outline of the scorpion depicted in the constellation of *Scorpius*. At the opposite end of the creature is Lambda Scorpii, otherwise *Shaula*, its sting.

Dubhe (star Alpha Ursae Majoris)
As in a number of constellations, the name of the Alpha star is an Arabic name for the constellation as a whole. Here it derives more precisely as a shortening of Ṭhahr al Dubb al Akbar, 'back of the greater bear', because that is its position midway along the back of the great bear who is *Ursa Major*.

Dumbbell Nebula (in *Vulpecula*)
The nebula, officially M 27 (in Messier's catalogue), is named for its shape.

Dusky Ring see **Crêpe Ring**

Eagle Nebula (in *Serpens*)
The Eagle Nebula, officially known as M 16 (NGC 6611), was discovered by the Swiss astronomer Jean-Philippe de Chéseaux in 1746, and is named for its fanciful resemblance to an eagle with outstretched wings.

Eagle, The see **Aquila**

Earth (planet)
Today the standard English word 'earth' has a number of senses, including 'ground', 'soil', 'burrow', 'electrical connection', and so on. As the name of the planet on which we live, it has been in use since the earliest times, since English was first recorded, so that it has also become more or less synonymous with 'world'. The classical names for the Earth also had other senses, so that Latin *Terra* also meant 'ground' or 'soil', as well as 'land' or 'country', and Greek *Ge* had similar senses. Today the name of the planet in many languages is thus the same as the word that means 'soil' or 'land', such as French *Terre*, German *Erde*, Russian *Zemlya*, and so on.

Eastern Sea see **Orientale, Mare**

Elara (satellite of *Jupiter*)
Elara is the seventh satellite of Jupiter, and was discovered by the American astronomer Charles Dillon Perrine in 1905, a year after he had discovered *Himalia*. All Jupiter's sixteen known satellites have names from classical mythology, and Elara was a somewhat obscure nymph who was the mother of Typheus by Zeus (Jupiter).

El Nath (star Beta Tauri)
Taurus is The Bull, and El Nath is 'the butting one', from Arabic Al Nāṭiḥ, because the star marks the tip of the bull's right (northern) horn, the other horn having Zeta Tauri at its tip. El Nath is thus at one end of the bull, at the greatest distance from the star that marks the animal's rear foot, Omega Tauri. Commenting on the extreme disposition of the two stars in the constellation, the English astronomer William Henry Smyth was prompted to ask, 'Can this have given rise to the otherwise pointless sarcasm of "not knowing B from a bull's foot"?'. El Nath was formerly regarded as being shared by two constellations: in Taurus it is (now officially) Beta Tauri, but in *Auriga* it was Gamma Aurigae, and when regarded as belonging to that constellation went by the Arabic name of Al Ka'b dhi'l 'Inān, 'the heel of the rein-holder', because it marked this point on the outline of the figure of The Charioteer. Compare the name of Beta Aurigae, *Menkalinan*.

Eltanin (star Gamma Draconis)
The star, which is the brightest in the constellation of *Draco*, has a name deriving from Arabic Al Rās al Tinnīn, 'the head of the dragon', because it is located in this part of the dragon in the night

sky. For the second part of the Arabic name, compare that of *Thuban*, the star Alpha Draconis. The reference here is to the feature (or part of it) that the star represents on the dragon who is the outline of the constellation *Draco*. With regard to the error in the name, see also *Alwaid*.

Enceladus (satellite of *Saturn*)

Enceladus is the second satellite of Saturn, and was discovered by William Herschel (the German-born English astronomer) in 1789, the same year that he discovered *Mimas*. He named both satellites after two of the giants who fought the gods in the 'Gigantomachy', or 'Battle of the giants'. Enceladus fought Athena, and was crushed to death by having the island of Sicily piled on top of him. As a group, the giants of Greek mythology were the offspring of Ge and *Uranus*, and so were brothers of Cronos (Saturn). Most of the planet's satellites were named after the Titans, the first-born children of Ge and Uranus.

Encke's Comet

The comet is unusually named not after its discoverer, but after the German astronomer who computed its return in 1822, Johann Franz Encke, after it had been initially discovered in 1786 by the French astronomer Pierre Méchain. Encke's Comet has the shortest known period of any, returning once every 3.3 years. See also *Encke's Division* and *Encke's Doodle*.

Encke's Division (in ring of *Saturn*)

The division in Ring A of *Saturn* (the outermost) was named after its discoverer, the German astronomer Johann Franz Encke.

Encke's Doodle (in *Encke's Division*)

Encke's Doodle (or the Encke Doodle) is a nickname given to an irregularly shaped 'kink' or ringlet in *Encke's Division* in Ring A of *Saturn*. It was discovered during the Saturn space probe made by Voyager 1 in 1980 and observed more closely by Voyager 2 subsequently.

Enif (star Epsilon Pegasi)

The name comes from Arabic Al Anf, 'the nose', because the star marks this particular organ on the outline of the winged horse depicted in *Pegasus*.

Epimetheus (satellite of *Saturn*)

Epimetheus was discovered by the French astronomer Audouin Dollfus in 1966 (during his observation of the newly discovered

Janus) and was subsequently named after the son of the Titan *Iapetus* in classical mythology.

Equuleus (constellation)
The name is an ancient one, usually translating into English as The Little Horse although also known as The Foal or The Colt. Because its neighbouring constellation is *Pegasus*, it may have been regarded as depicting that horse's brother, Celeris. The Greek name for it (which came first) was Hippou Protome, literally 'the horse's bust' (that is, its upper figure), and the Arabs called the constellation Al Kit'ah al Faras, 'the part of the mare'. With regard to this latter name, compare that of the constellation's brightest star, *Kitalpha* (Alpha Equulei). Equuleus is a small constellation and only a keen fancy can see a resemblance in its few stars to anything like a horse. See also *Pictor* for another 'horse'.

Eridanus (constellation)
Eridanus in classical mythology was the river into which Phaethon fell after his attempt to drive the chariot of his father Helios, the god of the sun. The present constellation of the name is usually known in English simply as The River, and can be seen in the sky as a long meandering chain of stars. (At one time it was even longer than it is now, and included the stars of *Fornax*.) The Greeks called it, as in English, simply 'the river' (ho Potamos), as did the Arabs (Al Nahr). Classicists and astronomers have debated for many years which precise river, if any, the Eridanus is supposed to represent, whether in mythology or in the night sky. Popular contenders have been the Po of Italy (probably the most likely), the Nile, or the Euphrates. For star names directly related to that of the river, see also *Acamar* and *Achernar*, and for more on the Arabic name, see the *Milky Way*.

Eros (asteroid 433)
Eros was discovered in 1898 by the German astronomer Carl Witt. In Greek mythology, Eros was the god of love, the son of Ares (*Mars*) and Aphrodite (*Venus*). Because the asteroid was the first known minor planet to come within the orbit of Mars, it was given a masculine name (like that of the Roman god), as distinct from the conventional asteroids, which have feminine names. But in mythology Eros is the Greek equivalent of *Amor*, an asteroid that also has an inter-Martian orbit, and the latter name was probably influenced by the former.

Er Rai (star Gamma Cephei)
The name represents Arabic, Al Rai͑, 'the shepherd'. This has no relation to the king who is depicted in the constellation of *Cepheus*, but is a purely Arabic name, because the Arabs knew the constellation, or at least many of its stars, as Al Aghnam, 'the sheep', as well as by a name corresponding to the present English one. See also the name of the star *Alfirk* (Beta Cephei).

Eskimo Nebula see **Clown Face Nebula**

Europa (1) (satellite of *Jupiter*)
Europa is the second *Galilean* satellite of Jupiter, discovered in 1610, and was named by its fellow discoverer, the German astronomer Simon Marius, after the girl who became the mistress of Zeus (Jupiter) in Greek mythology. Marius also named the satellites *Callisto*, *Ganymede* and *Io*, who are mythological characters also connected in some way with Jupiter.

Europa (2) (asteroid 52)
The asteroid was discovered in 1858 by the German astronomer Hermann Goldschmidt, and was named not after the Europa of mythology (see above) but for the continent of Europe. (Asteroid 67 was named Asia, for similar reasons.) In general, astronomical names do not duplicate, but here is an exception.

Evening Star, The see **Venus**

False Cross, The (group of stars in *Carina* and *Vela*)
The group comprises four stars, Iota and Epsilon Carinae and Kappa and Delta Velorum. They form a figure that somewhat resembles the *Southern Cross* and that has sometimes been mistaken for it by astronomers and navigators.

Fecunditatis, Mare ('sea' on Moon)
The name of the *mare* is usually translated in English as Sea of Fertility, and is one of the 'emotional' names such as *Mare Crisium* given by the Italian astronomer Giovanni Riccioli in his 'Almagestum Novum' of 1651. The favourable name is similar to those of the *Mare Tranquillitatis* and *Mare Serenitatis*, and in fact all three *maria* are more or less contiguous in the east central section of the near side, not far from the equally fruitful-sounding *Mare Imbrium* and *Mare Nectaris*.

Felis (former constellation)
There are dogs and horses in the night sky, so why not a cat? Felis

(The Cat) was a constellation introduced in 1805 by the French astronomer Joseph Jérôme Lalande, using stars between *Antlia* and *Hydra*. He justified his choice of name quite candidly: 'I like cats. I will let this figure scratch on the chart. The starry sky has worried me quite enough in my life, so that now I can have my joke with it.' One or two other astronomers included the new constellation in their catalogues and charts for a while, but the name did not catch on and by the end of the century had fallen quite into disuse. Truth to tell, the stars here could hardly be said to suggest any feline features anyway.

Fertility, Sea of see **Fecunditatis, Mare**

Fishes, The see **Pisces**

Fish Mouth, The (area in *Orion*)
The Fish Mouth, so named for its shape, is a dark-coloured stretch of gas between the two nebulae M 42 and M 43 (NBC 1976 and NBC 1982).

Flora (asteroid 8)
Flora was discovered in 1847 by the English astronomer John Russell Hind, the same year that he discovered *Iris*. It was named after the Roman goddess of flowering plants in classical mythology, and was the last of the asteroids in the first ten to be named after a Roman (as distinct from a Greek) goddess.

Flying Fish, The see **Volans**

Flying Horse, The see **Pegasus**

Foal, The see **Equuleus**

Fomalhaut (star Alpha Piscis Australis)
The French-looking name is actually Arabic, coming from Fum al Ḥūt, 'mouth of the fish'. The constellation of *Piscis Australis* represents a drinking fish, and Fomalhaut, its brightest star, is located at its mouth. It is the only named star in the constellation.

Fornax (constellation)
The constellation, whose Latin name means The Furnace, was introduced in the mid-eighteenth century by the French astronomer Nicolas Louis de Lacaille, originally with the full name of Fornax Chemica (or Fornax Chymiae), The Chemical Furnace. The name is typical of those chosen by Lacaille to represent the scientific and technical achievements and discoveries of his day. By 'furnace' here

he meant the chemical apparatus used in a laboratory for various heating operations and experiments.

Fox, The see **Vulpecula**

Frederici Honores (former constellation)
This was the name of a constellation introduced in 1787 by the German astronomer Johann Bode, and it means Frederick's Glory, in honour of Frederick II of Prussia (better known as Frederick the Great) who had died the previous year. It was located in the region between the constellations of *Cepheus*, *Andromeda*, *Cassiopeia* and *Cygnus* where, around a hundred years earlier, Augustin Royer had attempted to replace Hevelius's *Lacerta* by his *Sceptre and Hand of Justice*. By the end of the nineteenth century, however, both this latter constellation name and Frederici Honores had disappeared from the astronomers' charts and catalogues, with Lacerta alone prevailing.

Frederick's Glory see **Frederici Honores**

Frigoris, Mare ('sea' on Moon)
The name is translated in English as Sea of Cold, and is an irregularly shaped *mare* in the north central section of the near side of the Moon. It was introduced by the Italian astronomer Giovanni Riccioli in his lunar map of 1651, in turn based on names created by his pupil Grimaldi, and was doubtless chiefly suggested by the location of the region near the Moon's north pole. At the same time, Grimaldi liked contrasting names, and he may well have designed it to be a counterpart to his 'Terra Caloris', or 'Land of Heat', which no longer exists. Before Riccioli, the same 'sea' had appeared on the chart of the Flemish cosmographer Michel Langren as 'Mare Astrologorum', that is, 'Sea of Astronomers'.

Furnace, The see **Fornax**

Gabi (asteroid 1665)
As an example of a thoroughly modern astronomical name, by contrast with the many historic ones, here is Gabi, so named in 1970 after Gabriele Seyfert, the East German women's world ice-skating champion that year. The name was given in the familiar form not simply because the sports star was popularly known as Gabi, rather than Gabriele, but because Gabriella already existed for asteroid 355 (discovered in the 1890s).

Galileans (satellites of *Jupiter*)
The Galilean satellites are the four largest of Jupiter, *Io*, *Europa*,

Ganymede and *Callisto*, and are so called because they were first observed in detail by Galileo in 1610, in the pioneering days of observation by telescope. See also *Ganymede*.

Ganymede (satellite of *Jupiter*)
The *Galileans* were all named by Galileo's co-discoverer and observer, the German astronomer Simon Marius. All four satellites have names that in some way link them with Jupiter (Zeus). In Greek mythology, Ganymede was the handsome youth who, because of his beauty, was carried up to the heavens to be Zeus's cupbearer. For the stories and links behind the other three satellites see their respective entries.

Garnet Star, The (Mu Cephei)
The star Mu Cephei, which has no individual name, was called the Garnet Star by William Herschel, the English astronomer, because of its striking red colour. Technically it is a red supergiant, and the prototype of variable stars known as 'semiregular variables', that is, pulsating stars, although some astronomers regard it as irregular. (The brightest member of this category is *Betelgeuse*.)

Gassendi (crater on Moon)
The crater is on the north edge of the *Mare Humorum*, and was named after the seventeenth-century French astronomer Pierre Gassendi, better known today as a philosopher and mathematician.

Gaussia (asteroid 1001)
The asteroid, discovered by Soviet astronomers in the 1920s, was named after the nineteenth-century German astronomer Karl Gauss (who also gave his name to the magnetic unit known as the 'gauss'). There is a largish group of asteroids named after astronomers but with the feminine ending '-ia'. Examples of others are *Pickeringia* (784) and *Piazzia* (1000). Gauss played an important role in asteroidology, because his calculation led to the 'recovery' of *Ceres* a year after Piazzi had first discovered it in 1801.

Gemini (constellation)
The age-old constellation, whose Latin name means The Twins, represents a pair of twins holding hands. They are familiar as *Castor* and *Pollux*, the sons of Leda in classical mythology, and members of the Argonauts' crew. As such, they were the patron saints of sailors and navigators, and even today the two stars of Castor and Pollux (respectively Alpha and Beta Geminorum) are a reliable guide for measuring an angle, because they are known to be 4.5°

apart. The Greek name for the constellation, with the same meaning, was Didymoi, and the Arabic name of Al Tau'amān also meant 'the twins'. As the outline of the stars can fairly easily be thought of as depicting two figures holding hands, it is not surprising that they have been named over the centuries as other than the classical pair of Castor and Pollux. Among other starry pairs there have been David and Jonathan, Adam and Eve, and the two Egyptian gods Horus the Younger and Horus the Elder.

Geminids (meteor shower)
The annual meteor shower, occurring in mid-December, radiates from the constellation of *Gemini* and is named after it, as other meteor showers are named after their own constellations.

Gemma see **Alphecca**

Geographos (asteroid 1620)
The asteroid was named as a tribute to the National Geographic Society (of the USA), which in 1950 subsidised the photographic sky survey being undertaken by the Palomar Observatory in California.

Ghost of Jupiter, The (nebula in *Hydra*)
The planetary nebula, officially catalogued as NGC 5242, has a size that appears similar to that of the disk of *Jupiter*, hence its name.

Giacobinids (meteor shower)
The irregularly occurring meteor shower is also known as the Draconids, because it radiates from the constellation of *Draco*. However, the shower is chiefly associated with its parent, the Giacobini-Zinner Comet, which was first observed in 1900 by the Italian astronomer M. Giacobini (and again in 1913 by the German astronomer E. Zinner). The Giacobinids displayed well in 1933 and 1945.

Giedi see **Algedi**

Gienah (star Epsilon Cygni)
Cygnus is The Swan, and Gienah is its wing, and this is what the name means in Arabic (more precisely Al Janah, 'the wing'). For other parts of the bird's body see *Deneb* and *Sadr*.

Giraffe, The see **Camelopardalis**

Goat, The see **Capricornus**

Goldfish, The see **Dorado**

Gomeisa (star Beta Canis Minoris)
The name is a corruption of one of the Arabic names for the constellation of *Canis Minor* as a whole. This may have been Al Jummaizā, 'the sycamine' (or mulberry tree) or Al Ghumaiṣā´, 'the weeping one', with the latter referring perhaps to some legend. Some astronomers, however, have seen the name as deriving from Al Ghamūs, 'the puppy', but this was probably a translation of the constellation's Latin name.

Gould's Belt (belt of stars in *Orion*, etc)
The belt of bright stars, found not only in Orion but also in other constellations such as *Scorpius*, *Carina* and *Centaurus*, among others, was first pointed out in the nineteenth century by the English astronomer Sir John Herschel and was named after the American astronomer who made a special study of it, Benjamin Apthorp Gould.

Graffias (star Beta Scorpii)
The name is generally believed to derive from some word meaning 'crab', and so to be either synonymous with the name of the constellation of *Scorpius* as a whole or to denote the star at the base of one of the scorpion's claws.

Great Bear, The see **Ursa Major**

Greater Dog, The see **Canis Major**

Grimaldi (crater on Moon)
The crater, in the west central section of the near side of the Moon, is named after the Italian astronomer Francesco Grimaldi, a pupil of the famous Giovanni Riccioli in the seventeenth century. Riccioli based his star map of 1651 on the observations of his pupil, and used many of the names that Grimaldi himself placed on the Moon. Fittingly, Riccioli has his own crater next to that of his pupil, even though both are near the western limb (edge).

Grus (constellation)
The name is not an ancient one, but was introduced in 1603 by the German astronomer Johann Bayer. The Latin name is usually translated in English as The Crane, and the stars do seem to represent the outline of some large water bird. However, some observers have seen a different bird, Phoenicopterus, The Flamingo!

Gum Nebula, The (in *Vela*)
Unlike many nebulae, the Gum Nebula is not named for its resem-

blance to a particular creature or object, but after the Australian astronomer who first drew attention to it in 1952, Colin S. Gum. It is believed to be the remnant of one or more supernovae (see Glossary).

Hadar see **Agena**

Halawe (asteroid 518)
The asteroid was discovered in 1903 by the American astronomer R. S. Dugan, who gave it a more worldly name than many of its predecessors, because it was that of his favourite sweet, the Arabic confection made of crushed sesame seeds and honey (usually known in English as 'halva').

Halley's Comet
The best-known comet to the lay observer or non-astronomer, and one whose return can be confidently predicted every seventy-six years (last, rather disappointingly, in 1985). The comet itself obviously goes back hundreds if not thousands of years (one famous depiction of it is in the Bayeux Tapestry, because it was visible in 1066, just before the Norman Conquest), but it is named after the astronomer who first predicted its periodicity after observing it in 1682, Edmond Halley, England's second Astronomer Royal. It duly reappeared (after his death) in 1758.

Hamal (star Alpha Arietis)
Aries is The Ram, and the constellation's most prominent star, as frequently happens, has a name that is virtually identical to that of the whole group. Its full Arabic name was Al Rās al Hamal, 'the head of the sheep', and indeed its position in the outline of the animal is at its forehead. Some astronomers of the past also knew it as *El Nath*, although today this name is that of another star in another constellation, even though not inappropriate for Hamal.

Hare, The see **Lepus**

Haris (star Gamma Boötis)
The name means 'the keeper' (Arabic Al Hāris), perhaps as if the star was regarded as 'keeping an eye on' the other members of the group that together form *Boötes*, The Herdsman. An alternative name for the star is Seginus. The origin of this is uncertain. Some claim that it is a corrupt form of *Cepheus*, the name of the neighbouring constellation. Others see it as a distortion of the name of the constellation's brightest star, *Arcturus*, possibly via the Arabic version of this, Kheturus.

Hebe (asteroid 6)
Hebe was the sixth asteroid to be discovered, by the German astronomer Karl Ludwig Hencke in 1847. It was named after the Greek goddess of youth in classical mythology, who brought nectar and ambrosia to the gods on Olympus. Two years previously Hencke had discovered *Astraea*.

Hector (asteroid 624)
Hector was discovered in 1907 by the German astronomer August Kopff. It was given the name of the leader of the Trojan forces in the Trojan War. And as Hector was found as one of the *Trojans* in the orbit of *Jupiter*, it was natural that it should have been given this particular name, because in the classical story, Hector killed *Patroclus* (the asteroid discovered by Kopff the previous year) and was himself killed by *Achilles*, the asteroid that was the first of the Trojans to be discovered (in 1906).

Helix Nebula (in *Aquarius*)
The large but faint planetary nebula, with official number NGC 7293, is named for its shape, a circular patch.

Hercules (constellation)
The stars that form the outline of Hercules, the great superman of classical mythology, represent him as a kneeling figure, with one foot on the head of the dragon whom he killed in his Eleventh Labour, that is, the stars *Rastaban* and *Eltanin* in the neighbouring constellation of *Draco*. His kneeling posture is reflected in the name of Alpha Herculis, *Ras Algethi*, and in fact many of the early Greek names of the constellation indicated this specifically, and included for instance Engonasi (explained by Hipparchos as 'ho en gonasi kathemenos', 'bending on the knees') and Oklazon, 'the kneeler'. These names did not, however, identify the figure as that of Hercules, only as that of a kneeling man. The Romans translated these 'kneeling' names with such designations as Genuflexus, Ingeniculatus and Procidens. The equivalent Arabic name was Al Jāthiyy a'la Rukbataihi, 'the one who kneels on both knees'. Hardly surprisingly, this long name was eventually eroded and abbreviated to simply Elhathi. But the Romans also knew the figure as Saltator, 'the leaper', and this similarly had its Arabic equivalent, which was Al Raḳīs, 'the dancer'. This name was also one of the bynames of Hercules in classical mythology, so identified the hero more specifically.

Herdsman, The see **Boötes**

Hermes (asteroid)

The asteroid was detected in 1937 by the German astronomer Karl Wilhelm Reinmuth, who had earlier discovered *Apollo*. Both asteroids had distinctive properties (Hermes approached closer to the Earth than any other celestial body except the Moon), and as mentioned in Appendix II the names of some male mythological characters came to be used for asteroids that were in some way 'different' from the norm. Hermes was the Greek messenger of the gods. The asteroid has no official number because it was insufficiently observed, and it is now lost.

Herschel-Rigollet Comet

The comet was discovered in 1788 by Caroline Herschel, sister of the great English (German-born) astronomer, Sir William Herschel. It has a period of 156 years, and was recovered in 1939 by the French astronomer Robert Rigollet.

Hesperus see Venus

Hidalgo (asteroid 944)

The asteroid was discovered in 1920 by the German astronomer Walter Baade, who twenty years later was to discover *Icarus* also. It has a very eccentric orbit and the longest period (fourteen years) of any asteroid except *Chiron*. These special features required a masculine name for it (see *Hermes*) and Hidalgo is the title of a Spanish gentleman.

Himalia (satellite of *Jupiter*)

As Jupiter's sixth satellite, Himalia has the name of a character that, like most of the planet's other satellites, can be found to have some connection with Zeus (Jupiter) in classical mythology. In her case, she was a nymph of Rhodes with whom Zeus had intercourse after his victory over the *Titans*.

Hind's Crimson Star (R Leporis)

The star, in *Lepus* (and with R marking it as a variable, see Glossary), is named after the English astronomer John Russell Hind, who described it in 1845 as 'like a drop of blood on a black field'. Its colour is indeed intensely red.

Hipparchus (crater on Moon)

The crater, near the centre of the near side of the Moon, was named after the famous Greek astronomer Hipparchus. Many other famous Greek men of learning have similarly had craters and other features named in their honour, such as Anaxarogas, Aristotle, Archimedes, Pythagoras and Plutarch.

Homam (star Zeta Pegasi)
The Arabic name was originally Saʿd al Humām, 'lucky one of the hero', although it is not clear who the hero is. Presumably, at any rate in classical mythology, it is not *Pegasus* himself. Perhaps it is *Perseus*, because Pegasus was born from the blood of Medusa after Perseus had slain her. But more likely, and assuming the Arabic origin is correct, the precise reference has now been lost.

Horologium (constellation)
The name is usually rendered in English as either The Clock or The Pendulum Clock. The constellation was one of those with names of mechanical instruments introduced in the mid-eighteenth century by the French astronomer Nicolas Louis de Lacaille, and in common with most of his other introductions is a somewhat small and remote celestial region. Any resemblance of the outline to a clock with a pendulum, too, is more fanciful than factual.

Horsehead Nebula (in *Orion*)
The nebula is a dark cloud of dust (in an area of glowing gas) which certainly does suggest a horse's head, especially that of the knight in a game of chess. The official designation of the nebulous area of gas is NGC 2024. (Some astronomers prefer to call the nebula the Horse's Head, rather than the Horsehead.)

Horseshoe Nebula see **Omega Nebula**

Hubble's Variable Nebula (in *Monoceros*)
The small and faint nebula officially called NGC 2261 is named after the American astronomer Edwin Powell Hubble, who in the 1920s first established that some of the nebulae are really galaxies of gas, dust and stars far beyond our own Galaxy. The Nebula named after him contains a well-known variable star (R Mon, that is, a variable in Monoceros). Hubble also gave his name to a number of other astronomical terms, such as Hubble's Constant (relating to the velocities at which the galaxies are apparently receding) and Hubble Time (the time that has elapsed since the origin of the universe, tentatively put at somewhere between fifteen and twenty million years ago).

Humboldtianum, Mare ('sea' on Moon)
The *mare* (see Glossary) lies in the north-east section of the near side of the Moon and is named after the famous German scientist

and explorer of the eighteenth and nineteenth centuries, Alexander von Humboldt. He also gave his name to the Humboldt River in Nevada, USA, as well as a number of other natural and man-made geographical objects in that country and elsewhere.

Humboldt's Sea see **Humboldtianum, Mare**

Humorum, Mare ('sea' on Moon)
The Latin name of the 'sea' or *mare*, in the south-west section of the near side of the Moon, is usually translated as Sea of Moisture or Sea of Humours, although as is now well known the 'seas' of the Moon have never contained any water. The name first appeared on Grimaldi's map of the Moon, which was used by Riccioli in his own map (the 'Almagestum Novum') of 1651.

Hunter, The see **Orion**

Hyades (star cluster in *Taurus*)
In the constellation of Taurus, the Hyades are a V-shaped group of stars which appear as the bull's face. As such, they have been well known since ancient times, and their name is that of the daughters of Atlas and a nymph named either Aethra or Pleione. In classical mythology, they were said to have been formed into a constellation by Zeus, and their appearance in the sky coincided with the season of spring rain, hence the link with Greek *hyein*, 'to rain'. The popular Roman name for the stars, however, was Suculae, meaning 'little pigs', as if deriving from Latin *sus*, 'sow', translating (and related to) Greek *hys*. (Pliny combined both senses by explaining that the rains with which the stars were associated made the roads so miry and muddy that they seemed to delight in dirt, like pigs!) The most common Arabic name for the cluster was Al Debaran, 'the follower', since they seemed to follow the *Pleiades* (who were also the daughters of Atlas in classical mythology and who are also located as stars in Taurus). This Arabic name is the basis of the name of *Aldebaran*, and this star appears to be one of the Hyades but is actually distinct from them, as an unrelated star in the foreground.

Hydra (constellation)
Hydra is the largest constellation in the sky, with its name translating as The Water Snake. It is a faint constellation, however, and observers will need a telescope or at least binoculars to trace most of the stars in its sinuous body as it winds its way from the constellation of *Cancer*, in the northern hemisphere, down to the tip of

its tail south of the celestial equator by *Libra* and *Centaurus*. (Apart from *Alphard*, which marks the beast's heart, its brightest stars are those that form its head.) In classical mythology, Hydra is now usually identified with the many-headed monster (the Lernaean Hydra) that Hercules killed in his Second Labour. It has also been thought of, however, as some other snake or serpent, or even as a dragon, although this last creature is now the one reserved for the constellation of *Draco*. Properly it should now be thought of as the 'northern water snake', because *Hydrus* is its southern counterpart, introduced later. The Arabic name for Hydra was either Al Shujā͑ or Al Ḥayyah, both meaning 'the snake'.

Hydrus (constellation)
As mentioned for *Hydra*, Hydrus is the southern counterpart of the older, larger constellation. It was introduced in 1603 by the German astronomer Johann Bayer, and has a name that is usually translated into English as The Lesser Water Snake. (Grammatically the only difference between the two is not one of size but of gender, because Hydra as a Latin name is feminine, but Hydrus is masculine.) If anything, Hydrus looks less like a water snake than Hydra does, but its head could be said to be down by the neighbouring constellation of *Octans*, near the south celestial pole, and its tail up almost as far as *Achernar*, in neighbouring *Eridanus*. Hydrus has no named stars, and in point of fact is rather characterless, both as constellation and water snake.

Hygeia (asteroid 10)
Hygeia was the tenth asteroid to be discovered, in 1849, and it was the Italian astronomer Annibale de Gasparis, who was the discoverer. (He would discover several more during the course of the next few years.) The asteroid was given a name that ends a run of goddesses in Greek and Roman mythology, with Hygeia being the personification of health (of hygiene, in fact). See *Parthenope* (asteroid 11) for a breakaway from the names of goddesses, as well as for a little more relevant information about de Gasparis.

Hyperion (satellite of *Saturn*)
Hyperion is Saturn's seventh satellite, detected by the American astronomer G. P. Bond in 1848 as a 'star of the 17th magnitude in the plane of Saturn's ring, between Titan and Japetus' (see *Titan* and *Iapetus*). Bond was not the namer of the new satellite; that privilege fell to the amateur English astronomer William Lassell who had independently found the satellite before news of its

discovery had reached England. He proposed the name Hyperion as that of the mythological son of *Uranus* and Ge who was not only a *Titan* but also the sun-god. Lassell had thus reverted to the type of name that had previously been interrupted by Herschel when he named *Mimas* and *Enceladus* (in 1789) after Giants (as distinct from *Titans*, which see for the distinction). In 1898 the satellite *Phoebe* would be given a name that in turn was carefully contrasted with that of Hyperion.

Iapetus (satellite of *Saturn*)
Iapetus is the eighth satellite of Saturn, and was the second to be discovered, by the Italian astronomer Giovanni Domenico Cassini in 1671. Because the more or less generic mythological name *Titan* already existed for the first satellite of Saturn to be discovered (in 1655), Cassini chose a more 'concrete' name for his own discovery, so that Iapetus, himself a Titan, was the son of *Uranus* and Ge and one of the elder brothers of Cronos (Saturn).

Icarus (asteroid 1566)
Icarus was discovered in 1949 by Walter Baade, the German astronomer who also discovered *Hidalgo*. In classical mythology, Icarus was the rash youth who flew too near the sun so that the wax fastening his wings to his shoulders melted in the heat, and he plunged to earth to be drowned in the sea. The name is apt for the asteroid, which is one of the few known to have come closest to the Sun within the orbit of *Mercury*. (Another is Phaethon.)

Imbrium, Mare ('sea' on Moon)
The Latin name means Sea of Rains, and the *mare* (see Glossary) is located in the north central section of the near side of the Moon, where it can be found (with the naked eye) leading into the *Oceanus Procellarum* (Ocean of Storms) bounded by the *Apennines* and the Alps, a suitable site for a region with such a 'watery' name. It was one of the names that first appeared on Grimaldi's map and that was subsequently printed in Riccioli's 'Almagestum Novum' of 1651.

Indian, The see **Indus**

Indus (constellation)
The constellation, whose name translates as The Indian, was introduced in 1603 by the German astronomer Johann Bayer in his star atlas. Somewhat unexpectedly he regarded its stars as depicting an American ('Red') Indian, as a welcome change from the mythological names and mechanical instruments. Apparently Bayer saw

the outline as representing a figure carrying arrows in both hands, but no bow. It contains no individually named stars, and most of its stars are quite faint.

Ingenii, Mare ('sea' on Moon)
The 'sea' is one of the only two on the far side of the Moon (together with the *Mare Moscoviense*), where, however, it hardly qualifies as a genuine *mare* at all, because it is simply an irregularly shaped darkish area. Its name translates as Sea of Ingenuity, and is obviously a modern one, dating from the 1950s (after the Soviet space probe Luna 3 had photographed the far side for the first time), and referring to the cleverness of the Soviet scientists in managing this particular exploit. Many modern maps omit the name, although it usually appears on United States Air Force lunar charts.

Ingenuity, Sea of see **Ingenii, Mare**

Io (satellite of *Jupiter*)
Io is the innermost of the *Galilean* satellites of Jupiter, and was discovered by Galileo in 1610 (together with *Europa*, *Ganymede* and *Callisto*). Its name was proposed by the satellite's co-discoverer Simon Marius, with Io being the maiden loved by Zeus (Jupiter) and turned into a white heifer by him. (Compare *Europa*, another of Zeus's 'mistresses', with Zeus himself on this occasion turning into a white heifer.)

Iridum, Sinus (bay on Moon)
The Latin name means Bay of Rainbows, and designates the bay that leads out of the *Mare Imbrium*. It is typical of the many 'watery' names given to prominent features on the Moon in the sixteenth century.

Iris (asteroid 7)
The asteroid was discovered by the English astronomer John Russell Hind in 1847, when he also discovered *Flora*. The name is typical of the early asteroidal discoveries, being given after a goddess in classical mythology. Iris personified the rainbow and was a messenger (via the rainbow) between heaven and earth and gods and men.

Ishtar Terra (upland on *Venus*)
Ishtar Terra is one of the two main highland areas on Venus, the other being the larger *Aphrodite Terra*. Both regions bear the names of goddesses who are equated with the Roman Venus. Ishtar,

however, was not a personage in classical mythology, but was the Babylonian goddess of love and war. (The Greek version of her name was Astarte.)

Izar (star Epsilon Boötis)
The name is Arabic for 'girdle', because this is what it was regarded as representing on the figure of the herdsman (or hunter) who is depicted by the constellation of *Boötes*. The Arabs also knew the star by the longer name of Al Minṭakah al ʿAwwā', 'the belt of the shouter'. (Compare the name of *Alnitak*, the 'girdle' of *Orion*, as well as that of *Mizar* in *Ursa Major*.)

Janus (satellite of *Saturn*)
The satellite was discovered in 1966 on photographs taken by (among others) the French astronomer Audouin Dollfus. He named it Janus, rather unusually since he broke the run of Greek mythological names and also chose the name of a character who has no direct connection with Saturn. However, it could be argued that although the Roman god Janus was basically the god of doors and beginnings (hence the name of January), there are some legends which also tell how he received Saturn when he was driven from Greece by his son *Jupiter*, so that there was some kind of narrative link. (This particular account tells how Janus ruled a city on a hill called Janiculum, while Saturn ruled a city on a similar hill called Saturnia.)

Jewel Box, The (star cluster in *Crux*)
The name is an attractive if fairly obvious one for the star cluster, visible to the naked eye, that centres on Kappa Crucis. The cluster contains over fifty stars of different colours, for the most part white and blue, with one prominent red supergiant, and it was given its nickname by the English astronomer John Herschel in the nineteenth century.

Job's Coffin see **Delphinus**

Johanna (asteroid 127)
The asteroid was discovered in 1872 by the French astronomer Paul Henry, and was named patriotically in honour of Joan of Arc after the fall of the French Second Empire as a consequence of the Franco-Prussian war the previous year. The asteroid was given a German form of Joan's name because at that time the main research centre for asteroids was in Berlin.

Juno (asteroid 3)

Juno was the third asteroid to be discovered, and its finder, in 1804, was the German astronomer Karl Ludwig Harding, a member of the so-called 'Celestial Police' (the band of astronomers who met together in 1800 to search for a new planet between *Mars* and *Jupiter* but who found asteroids instead). The original name proposed for Juno was Hera, but the asteroidal expert (and member of the 'Celestial Police') Heinrich Olbers maintained that the asteroids should have the names of Roman, not Greek, goddesses because the first asteroid to be named, *Ceres*, had a Roman name. Thus the asteroid was endowed with the name of the Roman goddess of marriage, who corresponded to the Greek Hera. The name Hera was, however, subsequently given to an asteroid exactly one hundred numbered asteroids later (103), while asteroid 146 was named as Lucina, the byname of Juno.

Jupiter (planet)

Jupiter is the largest of the planets, and has long been a conspicuous and bright celestial object from ancient times, and for its prominence was hardly surprisingly given the name of the king and ruler of the Olympian gods in classical mythology. The Greeks originally knew the planet as Phaethon, meaning 'shining' or 'radiant', because by night it was the brightest of the planets when *Venus* was not visible, while in the fourth century BC Aristotle named Jupiter more concretely as Zeus. The Romans then renamed the planet by the name of their corresponding god, Jupiter, as they did with the other ancient planets of *Mercury*, *Venus*, *Mars* and *Saturn*. As a planet, Jupiter has an extensive family of satellites in the solar system, and almost all of them have names that relate directly to Zeus (Jupiter) in mythology. Four of them (*Io, Europa, Ganymede* and *Callisto*) were in fact the first satellites of any planet other than the Earth to be discovered, and when Galileo did so in 1610 through his famous newly improved astronomical telescope he named them collectively as the Sidera Medicea, 'Medican stars', in honour of his former pupil and future protector, Cosimo de' Medici, Grand Duke of Tuscany. 'Sidera' was thus the original designation of satellites, and it was only the following year that Kepler first used the latter word in its present sense (to apply to these same satellites of Jupiter). See also *Galileans* as a term for these first four, and 'satellite' in the Glossary.

Kaffaljidhmah (star Gamma Ceti)

The name represents Arabic Al Kaff al Jidhmah, 'the palm of the hand', and was properly used by Arab astronomers to apply to all those stars that formed the head of *Cetus*, The Whale, that is, *Menkar* (Alpha Ceti), Delta, Lambda, Mu and Xi Ceti, and Kaffaljidhmah itself. Later the name came to be used solely of Gamma Ceti. The Arabic name basically represented the same figure, the 'stained hand', that they saw in the constellation of *Cassiopeia* (which see).

Kapteyn's Star (in *Pictor*)
The star, a red dwarf (see Glossary), was named after Jacobus Cornelius Kapteyn, the Dutch astronomer who discovered in 1897 that it had the greatest proper motion of any known star except for *Barnard's Star* (in *Ophiuchus*).

Kaus Australis (star Epsilon Sagittarii)
Sagittarius is The Archer, and Kaus Australis is the star that represents the southern part of his bow. The name is a hybrid, a combination of Arabic Kaus, 'bow', and Latin Australis, 'southern'. Compare *Kaus Borealis* and *Kaus Meridionalis* and see *Al Nasl* for the star that marks the point of The Archer's arrow.

Kaus Borealis (star Lambda Sagittarii)
Just as *Kaus Australis* marks the southern part of The Archer's bow, so Kaus Borealis represents its northern end. The name is a similar hybrid of Arabic and Latin, with Borealis meaning 'northern'. See also *Kaus Meridionalis*.

Kaus Meridionalis (star Delta Sagittarii)
Here is the middle of The Archer's bow in the constellation of *Sagittarius*. For the northern and southern parts of it, see *Kaus Australis* and *Kaus Borealis*. Meridionalis is simply Latin for 'middle'. Some astronomical maps show the name as Kaus Media.

Keel, The see **Carina**

Kepler's Star (in *Ophiuchus*)
The star is the last but one supernova (see Glossary) observed in our Galaxy, in 1604, and it was named after the German astronomer who studied it, Johannes Kepler. Many supernovae have since been observed in other galaxies, and in early 1987 a new supernova, the brightest since Kepler's Star, was observed to be blazing out in the Large *Magellanic Cloud* in our own Galaxy. Astronomers now eagerly await the next supernova (see *Keyhole Nebula* for a possible candidate).

Keyhole Nebula, The (in *Carina*)
The name refers to a kind of dark-coloured 'notch' that appears in the nebula surrounding the star Eta Carinae. The official designation of the nebula is NGC 3372. Eta Carinae is believed to be a candidate for a future supernova.

Kids, The (stars Epsilon, Zeta and Eta Aurigae)
Auriga is The Charioteer (or The Wagoner), although the Greeks identified Alpha Aurigae as the she-goat Amalthea in classical mythology (see *Capella*). If this concept is developed, then Epsilon, Zeta and Eta Aurigae are seen as the goat's kids. This name is an ancient one, and the Greeks knew the three stars as Eriphoi, 'the kids', while the Roman name for them, with the same meaning, was Haedi.

Kitalpha (star Alpha Equulei)
As with many of the older constellations, named by the Greeks or Romans, the most prominent star has a name that equally well applies to the grouping as a whole. In this case, for *Equuleus*, The Little Horse, Kitalpha is a condensed and corrupt version of Arabic Al Kit'ah al Faras, 'the part of a horse'. See *Equuleus*. (The second half of the name thus has no connection with the Greek letter Alpha, which designates it.)

Known Sea, The see **Cognitum, Mare**

Kochab (star Beta Ursae Minoris)
The name represents Arabic Al Kaukab al Shamāliyy, 'the star of the north', and also originally applied to Alpha Ursae Minoris (see *Polaris*, as the Pole Star). Both Kochab and *Pherkad* (Gamma Ursae Minoris) are sometimes known as 'the Guardians of the Pole', despite the relatively great distance of both stars (in the 'bowl' of the Little Dipper, as *Ursa Minor* is also called) from the celestial north pole and the Pole Star itself.

Kornephoros (star Beta Herculis)
The Greek name, properly Corynephoros, means 'club-bearer', and was one of the several alternative names that originally applied to the figure of *Hercules* in the constellation as a whole. Most of the other star names in the depiction refer to different parts of Hercules' body, such as *Ras Algethi* (his head), *Marfak* (his elbow) and *Maasim* (his wrist).

Lacerta (constellation)
The constellation was introduced in the seventeenth century by the

German astronomer Johannes Hevelius, and its Latin name is usually translated in English as The Lizard. This refers to the sinuous outline of the main stars in the group, as they zigzag southwards from Beta Lacertae. Earlier constellations in this same region were the *Sceptre and Hand of Justice* and *Frederici Honores*, but both these names were discarded in favour of Lacerta. Hevelius introduced six other constellation names in his time: *Canes Venatici*, *Leo Minor*, *Lynx*, *Scutum*, *Sextans* and *Vulpecula*. For more about his naming of features on the Moon, see the Introduction (p. 10). Lacerta contains no named stars.

Lagoon Nebula (in *Sagittarius*)
The Lagoon Nebula, which is (just about) visible to the naked eye, is so named for its dark channels or 'lanes', which give the appearance of a lagoon between the main star cluster and the smaller one (NGC 6530). The overall official designation of the Lagoon Nebula is M 8 (after Messier), or NGC 6523.

Lalande 21185 (star in *Ursa Major*)
The name is that of a faint red dwarf (see Glossary), so designated because it appeared with this number in a catalogue compiled by the eighteenth-century French astronomer Joseph Lalande.

Large Magellanic Cloud see **Magellanic Clouds**

La Superba (star in *Canes Venatici*)
The star is a deep red variable (see Glossary), and its Italian name, meaning 'the superb one', was given to it for its 'superb flashing brilliancy' by the nineteenth-century Italian astronomer, Father Angelo Secchi. Its official designation is Y Canum Venaticorum (Y being the labelling Roman letter, see Introduction, p. 19).

Leda (satellite of *Jupiter*)
The satellite was discovered photographically in 1974 by the American astronomer Charles Kowal, and was named after the Queen of Sparta in Greek mythology who gave birth to Helen of Troy after Zeus (Jupiter) had visited her as a swan.

Leo (constellation)
The name is an ancient one, with The Lion (as it is known in English) being the one that was slain by Hercules as his First Labour. For once, the figure does actually look like the animal so named, and it is not difficult to make out the depiction of a crouching lion, with *Regulus* (Alpha Leonis) one of the stars marking his head and *Denebola* (Beta Leonis) indicating his tail.

The Greek name was Leon, with identical meaning, and the Arabic, also designating this animal, was Asad. 'Lion', too, was the constellation's name in many other languages, including Persian, Turkish and Hebrew, and the Babylonians similarly knew it as such. The stars forming the lion's head have sometimes been called The Sickle, from the curving outline of the stars here, with Regulus representing the handle. The other five stars in this subgroup are thus (in order along the outline) Eta Leonis, *Algieba* (Gamma Leonis), Zeta Leonis, Mu Leonis and Epsilon Leonis.

Leo Minor (constellation)
The Lesser Lion, as it is usually known in English, cannot hold a candle to its splendid elder brother, *Leo*, and its name was one of those introduced in the seventeenth century by the German astronomer Johannes Hevelius. Its few stars, all unnamed, and some not even properly catalogued, bear little resemblance to a lion or indeed to any kind of animal. The constellation is, however, located between its bigger and better namesake and *Ursa Major*, so has some justification for its name.

Leonids (meteor shower)
As with similar meteor showers, the Leonids are named after the constellation from which they appear to radiate, in their case that of *Leo*, with their peak time of observation centring on mid-November. (Their actual point of radiation is close to the star *Algieba*, Gamma Leonis.)

Lepus (constellation)
Like *Leo*, Lepus is a constellation with an ancient name, with its English translation usually The Hare. It is not clear exactly which hare is depicted here, although aptly enough Lepus can be found at the feet of *Orion*, The Hunter, who liked hunting hares. Some astronymists (interpreters of star names) have tried to see a connection with the Moon, because the 'Man in the Moon' has on occasions been seen as a hare (or a rabbit). The Greek name for the constellation was Lagos, having the same meaning, while the Arabic name, also denoting the animal, was Al Arnab. (Compare the name of its principal star, *Arneb*, Alpha Leporis.) The outline of the stars does in some ways suggest a hare with four legs. But for quite another animal see *Nihal* (Beta Leporis).

Lesser Dog, The see Canis Minor

Libra (constellation)
At first sight, the constellation name might seem to belong to the

collection of 'mechanical instruments' introduced by the Frenchman Lacaille in the mid-eighteenth century, as Libra means The Scales (or The Balance). However, the name is actually an old one. Originally the Greeks knew it as Chelai, 'the claws', because they saw the stars as suggesting the claws of a scorpion, or more precisely as the claws of The Scorpion, that is, as an extension of the neighbouring constellation of *Scorpius*. However, the Romans saw it differently, and in the time of Julius Caesar called it Jugum or Libra, meaning respectively 'beam of a balance' and 'balance' itself. The latter name predominated, with the balance regarded as being the symbol of justice held by the goddess of Justice, Astraea. (A modern representation of this concept can be seen in the statue of Justice above the Central Criminal Court, or Old Bailey, in London, where the female figure holds a pair of scales.) Certainly the stars that comprise Libra do suggest a pendant balance, and possibly Astraea is supposed to be represented by the neighbouring constellation of *Virgo*. At the same time Libra's stars resemble a pair of crab's (or scorpion's) claws, and this alternative depiction is preserved in the names of its three main stars, *Zubenelgenubi*, *Zubeneschamali* and *Zubenelakrab*, respectively Alpha, Beta and Gamma Librae. The Arabic name for the constellation was thus Al Zubānā, 'the claws', representing the root word on which these three star names are based.

Lion, The see **Leo**

Little Bear, The see **Ursa Minor**

Little Dipper, The see **Ursa Minor**

Little Dog, The see **Canis Minor**

Little Horse, The see **Equuleus**

Little Lion, The see **Leo Minor**

Little Snake, The see **Hydrus**

Lizard, The see **Lacerta**

Lucifer see **Venus**

Lupus (constellation)
The Latin name means The Wolf, and although the constellation was known to the Greeks and Romans in classical times, their own names for it were less specific, and they simply called it 'the wild animal' (Greek Pherion, Latin Fera or Bestia). The wolf name

seems to have first appeared as Lupus in the so-called 'Alfonsine Tables' of the thirteenth century, the astronomical catalogue drawn up by Arabian or Moorish astronomers at Toledo under the patronage of the future King Alfonso X, of Castile and León. The Arabs, however, called the constellation Al Asadah, 'the lioness'. Whatever the animal, it is depicted as being held in the grip of neighbouring *Centaurus*. The classical astronomers would have seen this as representing an offering to the gods. Despite its many interesting objects, Lupus has no named stars, almost as if it were overshadowed by the *Milky Way* in which it lies, or 'squeezed' by its two high-power neighbours of *Scorpius* and Centaurus.

Lynx (constellation)
Although a large constellation, Lynx is an obscure one, and it was introduced in the late seventeenth century by the German astronomer Johannes Hevelius to fill a gap between *Ursa Major* and *Auriga*. He is said to have given the name not for any particular depiction of a lynx in the constellation's stars, but simply because an observer needs to be 'lynx-eyed' to spot it! However, the formation was also formerly known as Tigris, The Tiger, apparently because its many small stars suggest the spots on a tiger's coat. Lynx has no named stars, and in fact apart from its most prominent star, Alpha Lyncis, all its members are designated by so-called Flamsteed numbers, not Greek letters. These are based on a catalogue of nearly 3000 stars entitled 'Historia Coelestis Britannica' ('British History of the Heavens') compiled by the first British Astronomer Royal, John Flamsteed, who died in 1719. (The catalogue was published posthumously, in 1725.) Flamsteed himself listed the stars in each constellation in order of right ascension (see Glossary), although the numbers by which astronomers know the stars today were not given by him but were added later. Among the brightest stars in Lynx are thus 5 Lyncis, 12 Lyncis, 15 Lyncis, 19 Lyncis, and so on, many of them double stars.

Lyra (constellation)
Known as The Lyre in English, the constellation name dates back to classical times and represents the stringed instrument invented by Hermes, messenger of the gods, and given by Apollo to Orpheus. The configuration of the stars does suggest a lyre or ancient harp, and the Greek name for the constellation, Cithara, had the same sense as the more closely corresponding Lyra, with the Latin name taken direct from the Greek. The Arabic name was similarly Al Ṣanj. But the Arabs also saw certain stars in the

grouping as eagles, in particular Alpha, Epsilon and Zeta Lyrae, which they called Al Naṣr al Waḳiʿ, 'the swooping eagle', because these stars suggest a bird with its wings half-closed, while another name, Al Naṣr al Ṭāʾir, 'the flying eagle', included the smaller stars Beta and Gamma Lyrae, representing the bird's outstretched wings. Alpha Lyrae is *Vega*, which is a corruption of the second half of the first of these Arab names. For quite another creature, see the names of Beta and Gamma Lyrae, respectively *Sheliak* and *Sulafat*.

Lyre, The see **Lyra**

Lysithea (satellite of *Jupiter*)
Lysithea was discovered in 1938 by the American astronomer Seth Barnes Nicholson, who also discovered *Ananke*, *Carme* and *Sinope*. In common with the names of Jupiter's other satellites, Lysithea was a mythological character having a connection with Zeus (Jupiter). In her case, though, the link is a tenuous one, and in fact her name does not feature in many dictionaries of classical mythology. However, according to one account she was mentioned as the mother of the 'first Dionysus', born to Zeus by Persephone, so the choice of her name for Jupiter's tenth satellite is (just about) justified.

Maasim (star Lambda Herculis)
The name is Arabic, representing Miʿṣam, 'wrist', because this is the portion of the body of *Hercules* that the star was intended to denote. However, owing to an error on the part of the German astronomer Johann Bayer, the name came to be given to Lambda Herculis not far from the giant's left shoulder, whereas it should really apply to Omicron Herculis (as it did in Ptolemy's star maps, and those of other early astronomers). So Maasim is an anatomical misnomer!

Maffei Galaxies (in *Cassiopeia*)
The name applies to two galaxies discovered in 1968 by the Italian astronomer Paolo Maffei, and they are designated after him. As yet we know little about them, because they are greatly obscured.

Magellanic Clouds (in *Dorado* and *Tucana*)
The names of Large Magellanic Cloud and Small Magellanic Cloud are given to the two satellite galaxies that accompany our own *Milky Way*. (They are also sometimes known respectively as Nubeculae Major and Minor.) Early mariners and navigators knew the two formations as the 'Cape Clouds', as they both appear in the

southern hemisphere and were associated in some way with the Cape of Good Hope, either because they seemed to resemble it in outline, or (more probably) because they were conspicuous celestial objects to navigators approaching the Cape. They were subsequently renamed after Ferdinand Magellan (Fernão de Magalhães), the Portuguese explorer and navigator, who is said to have first described them during his round-the-world voyage of 1519–22 (in which he rounded the Cape of Good Hope). Navigators used the Magellanic Clouds to establish the location of the south (celestial) pole, because between them they mark two angles of an almost equilateral triangle, with the south pole forming the third angle.

Manger, The see **Praesepe**

Mare ('sea' on Moon)
For the name of an individual *mare*, see the next word (for example for *Mare Tranquillitatis* see *Tranquillitatis, Mare*). For the nature of a *mare*, see Glossary.

Marfak (star Kappa Herculis)
The name comes from Arabic Al Marfik, 'the elbow', since this is the part of the body of *Hercules* that the star represents. (The Arab astronomers usually named the giant's right elbow thus, although later astronomers have designated his left elbow with the name.) Exactly the same name is still sometimes used for the star Lambda Ophiuchi, where it similarly marks the elbow of the celestial figure (here that of *Ophiuchus*). See also *Mirfak*.

Marginis, Mare ('sea' on Moon)
The Latin name means Border Sea, and denotes the *mare* (see Glossary) at the eastern edge (limb) of the near side of the Moon. The name dates from the seventeenth century.

Marilyn (asteroid 1486)
The asteroid was discovered in the 1930s by the American astronomer Paul Herget, who named it after his daughter. Paul Herget was specially interested in the names of asteroids, and is the author of a book on the subject (see Bibliography).

Mariner's Compass, The see **Pyxis**

Markab (star Alpha Pegasi)
The star marks a key point in the outline of the winged horse that is represented in the constellation of *Pegasus*, and its Arabic name

means 'saddle', because this is what it indicates. Markab is one of
the four stars that mark the distinctive square seen in the outline.
The others are *Scheat*, *Algenib* and *Sirrah*, respectively Beta Pegasi,
Gamma Pegasi and Alpha Andromedae (but originally Delta
Pegasi).

Mars (planet)
The well-known planet, formerly regarded as a 'twin' of *Earth*, and
famous for its 'canals' and possible water and life, was named after
the Roman god of war (known to the Greeks as Ares), on account
of its prominent red colour. This same colour also lay behind the
Greek name of the planet, which was Pyroeis, 'fiery' or 'flaming'.
It was Aristotle who called it Ares in the fourth century BC,
following which the Romans named it Mars, after their god who
corresponded to Ares. But the association with war and destruction
goes back even earlier than the Greeks, and the Babylonians similarly called the planet after their god of death and pestilence, who
was Nergal. Mars was thus associated with earthly calamities and
catastrophes, its red colour standing as a symbol of bloodshed. See
also *Antares*. In the event, Mars has turned out disappointingly,
because it has been shown to have not only no canals, but also no
water and, as far as can be determined, no signs of life.

Matar (star Eta Pegasi)
The Arabic name probably derives from the full designation of Al
Saʻd al Maṭar, 'the good fortune of the rain', with this applied
jointly to both Eta and Omicron Pegasi. This accords with the
Arabic name of the famous 'four-star square' in the constellation,
which was known as Al Dalw, 'the water bucket', with Matar being
'poured' out of it as welcome rain. (See *Pegasus* itself for more on
the 'water' theme, and also *Markab*.)

Medii, Sinus ('bay' on Moon)
The 'bay' has a name that means Middle Bay, referring to its central
position on the near side of the Moon. This is thus one of the
few 'locational' names, like that of the *Mare Marginis* (which see
above).

Megrez (star Delta Ursae Majoris)
Several of the star names in the constellation of *Ursa Major* refer
to points on the bear's body. This is one of them, since it derives
from Arabic Al Maghrez, 'the root of the tail'. If the stars here are
more clearly seen as the Big Dipper, then Megrez marks the point

where the handle joins on to the dipper (or 'saucepan', as it is popularly seen by some).

Menkalinan (star Beta Aurigae)
Auriga is The Charioteer, and Menkalinan is his shoulder! The name, as one would expect, is Arabic in origin, deriving from Al Mankib dhi'l 'Inān, 'the shoulder of the rein-holder'. Compare the name of *El Nath*, when regarded as Gamma Aurigae, rather than Beta Tauri.

Menkar (star Alpha Ceti)
Cetus is The Whale, or The Sea Monster, and Menkar represents Arabic Al Minhar, 'the nose', because this is the feature on the creature that it is supposed to mark. In reality, however, it marks the monster's open jaws, not its nose.

Mensa (constellation)
The Latin name means not so much simply 'the table' as The Table Mountain, because it was introduced by the French astronomer Nicolas Louis de Lacaille in the 1750s to commemorate the Table Mountain at the Cape of Good Hope, in South Africa, from where he had made his survey of the southern skies. However, a portion of the Large Magellanic Cloud edges from neighbouring *Dorado* into Mensa, and this may additionally have suggested to Lacaille the cloud that hangs over the real Table Mountain. (See *Magellanic Clouds* for more about the role of the Cape of Good Hope in star-naming.) When originally named by Lacaille, the constellation was in full Mons Mensae, and it was more recent astronomers who adopted the one-word version.

Merak (star Beta Ursae Majoris)
The Great Bear who is the subject of the constellation of *Ursa Major* has various portions of his anatomy denoted by individual star names. In the case of Merak, it is his loin, with the name representing Arabic Al Marākk.

Mercury (planet)
The Greeks knew Mercury by three names. A general name at first for the planet was Stilbon, 'the shiner' or 'the bright one'. But they also regarded it as both a 'morning star' and an 'evening star', because it was best seen either low in the east before dawn, or low in the west after sunset. As the former apparition, it was called Apollo, after the god of light (among other things). As an 'evening star', however, they knew it as Hermes, after the messenger of the

gods whose Roman counterpart was Mercury. It was Aristotle who introduced the latter Greek name. Mercury was originally Stilbon, 'the shiner', because it always appeared to accompany the Sun, and so was regarded as a 'spark', a sort of 'fiery spin-off', of the Sun, which was itself seen as the great fire of the sky. Mercury is, after all, the planet that is nearest to the Sun in the Solar System. For a better-known 'morning star' and 'evening star' with a twice-daily appearance, see *Venus*.

Mesarthim (star Gamma Arietis)
If the name has an Arabic origin, it has been lost through corruption. Some early astronomers claimed that the name derives from a Hebrew word meaning 'ministers', but the significance of this is unclear. It is just possible the name is connected in some way with that of *Sheratan*, Beta Arietis, because the two stars were associated in the calendar.

Metis (asteroid 9)
The asteroid was discovered in 1848 and was nearly named Thetis, a name proposed by Sir John Herschel. However, Metis was chosen, and Thetis kept for planet 17. Metis was an Oceanid (a daughter of Oceanus and Tethys) and the first wife of Zeus.

Miaplacidus (star Beta Carinae)
The unusual name at first sight appears to be of Roman origin, and has baffled many astronomers and astronymists, including the assiduous Richard Hinckley Allen (see Bibliography). If the second half of the name actually *is* Latin *placidus*, 'calm', 'peaceful', then the first part should properly be a noun, although not necessarily a Latin one. It could conceivably be a form of Arabic *moi*, 'water', with the whole understood as something like 'calm sea'. This would then to an extent accord with the name of the constellation *Carina*, The Keel, and equally well with the original constellation from which Carina was later partitioned, that of *Argo*. Jason and the Argonauts would certainly have wished for calm waters when making their voyage, as any mariner would. However, this explanation of the name is purely a tentative suggestion, and it should be said that most of the star names in the original constellation of Argo are not noticeably 'watery'.

Microscope, The see Microscopium

Microscopium (constellation)
This Latin name means The Microscope and, as one expects, the

name is a comparatively recent one. It was one of those introduced in the mid-eighteenth century by the French astronomer Nicolas Louis de Lacaille. As with many of his new constellations, the name is much more impressive than the actual star composition, and Microscopium is noticeably small, even insignificant, as it lies sandwiched rather artificially between *Piscis Austrinus* and *Sagittarius*. However, it must be admitted that the angular outline of its stars could suggest a basic optical instrument such as a microscope, and one assumes that Lacaille was not perpetrating a pun in giving the name to a 'microscopic' constellation.

Milk Dipper, The (group of stars in *Sagittarius*)
The pleasant nickname for the ladle-shaped formation of stars in the constellation is also an apt one, as Sagittarius lies in the *Milky Way*, moreover, in a particularly 'rich' section of it starwise. The actual stars that comprise the Milk Dipper are respectively Zeta, Tau, Sigma, Phi and Lambda Sagittarii, with the last named (otherwise *Kaus Borealis*) forming the handle. So romantic or poetic stargazers can here 'dip' into the rich cream of the Milky Way and drink their fill. They may care to accompany their drink with the 'meat' of the next entry.

Milky Way, The (band of stars stretching across the night sky)
The definition here of the Milky Way is necessarily rather diffuse, as the phenomenon itself is. Moreover, as explained in the Introduction (p. 6) and in the 'galaxy' entry in the Glossary, the two names 'Milky Way' and 'Galaxy' are inextricably linked (and sometimes confused), with the former being the English equivalent of the latter word, which directly reflects the Greek. The Milky Way has had countless names in different languages since the earliest times. But perhaps it is with the more familiar Greeks that we should directly begin. In the stories of classical mythology, the origin of the Milky Way has its own legend. In this, Heracles, on the order of his father Zeus, was brought as a baby to the breast of the sleeping goddess Hera, so that she could suckle him and make him immortal. (He was not already immortal because his mother had been the 'mere mortal' Alcmene.) However, on awaking, Hera pushed the baby from her breast, with the result that the milk spurted out across the heavens and that *it* remained immortal (after all, it is still there), while Heracles remained mortal. A charming but naïve account, of course, to explain the origin of the celestial object, and doubtless one created specially for this particular purpose. However, the Greeks already knew the Milky

Way as simply Gala, 'milk', or Kyklos galaktikos, 'milky circle' (hence 'galaxy'), so that the story had an existing basis on which to develop. This same Greek name was translated by the Romans, who called it Circulus lacteus, while the Arabic name for the Milky Way was originally Al Nahr, 'the river', a name subsequently transferred to *Eridanus*. Another Roman name for the Milky Way, of which the English is a direct translation, was Via Lactea (or Via Lactis, 'way of milk'). However, the early Anglo-Saxons also called it Watling Street (or the equivalent of this in Old English), a name that today remains an earthly one for the Roman road that runs from south-east England to North Wales. Among other names (here translated) were the Scandinavians' Woden Way (or Odin's Way), the Turks' Pilgrims Way (thinking of their annual pilgrimage to Mecca), and the Chinese Yellow Road, comparing the scattered stars to strewn straw. Indeed, Straw Way was itself a common name for the galaxy over much of Asia and Africa.

Mimas (satellite of *Saturn*)
Mimas, the innermost of Saturn's satellites, was one of the two (the other was *Enceladus*) discovered in 1789 by William Herschel. He named both satellites after Giants (as distinct from Titans) in classical mythology, with Mimas being killed by Ares (or Hephaestus, or Zeus, depending which account one follows) in the 'Gigantomachy', or battle between the Giants and the gods.

Mintaka (star Delta Orionis)
Orion, The Hunter, is depicted in his constellation as a figure whose various accoutrements are marked by individual stars. Mintaka marks his belt, with the name deriving from Arabic Al Minṭakah, 'the belt'. For a similar name, see the alternative Arabic name of *Izar* (Epsilon Boötis).

Mira (star Omicron Ceti)
Mira (also called Mira Ceti) is a prototype long-period variable star (see Glossary) with a characteristic deep red colour. Its name is Latin for 'wonderful' (grammatically feminine, to imply Stella Mira), this not simply denoting its striking appearance as a 'red giant' and its changing brightness as a variable star, but also appropriate for the first variable to be discovered. The star was first recorded in 1596 by the Dutch astronomer David Fabricius, and it was first recognised as a variable by his compatriot, Johannes Phocylides Holwarda, in 1638.

Mirach (star Beta Andromedae)

Mirach represents the waist of *Andromeda*, as she stands chained to the rock, and its name derives as a corruption of Arabic Mi'zar, 'girdle' or 'cloth round the waist'. A later Arabic name for this same star was Al Janb al Musalsalah, 'the side of the chained woman'. This name envisages the star as marking Andromeda's right side, although Mirach is today shown as indicating her left hip. For an identical name with this meaning, see *Mizar*.

Miranda (satellite of *Uranus*)

Miranda is not only the innermost of Uranus's five named satellites, but also one of the most recent to be discovered (by the Dutch astronomer Gerard Peter Kuiper, in 1948). Earlier names had been chosen from Shakespeare, and in particular from his play *A Midsummer Night's Dream* (*Oberon* and *Titania*). Kuiper therefore had a precedent to follow, and he named his discovery after Miranda in *The Tempest*, where she is the daughter of Prospero. Uranus has several more smaller satellites, ten of which were located by Voyager 2 in 1986 (see Introduction, p. 23). For another of Kuiper's namings, see *Nereid*.

Mirfak (star Alpha Persei)

As distinct from its other name of *Algenib*, the star's present name means 'elbow', because this was what some Arabian astronomers regarded it as representing on the depiction of *Perseus*. Its fuller title was Marfik al Thurayya, 'elbow of the Pleiades', that is, the one that was nearer to the *Pleiades* than the other elbow. See also *Marfak*, an identical name (but of a star in *Hercules*).

Mirzam (star Beta Canis Majoris)

Mirzam is a blue giant (see Glossary) whose name derives from Arabic Al Murzim, 'the announcer', so called because the appearance of the star in the sky was regarded as 'announcing' the imminent rising of the even more prominent *Sirius*. In the representation of The Greater Dog who is the constellation *Canis Major*, Mirzam marks his front right foot.

Mizar (star Zeta Ursae Majoris)

The famous double star has a name that is identical to that of *Mirach*, deriving from Arabic Mi'zar, 'girdle'. However, there appears to have been some confusion or misunderstanding here, as Mizar appears halfway along the tail of The Great Bear who is *Ursa Major*, where it can hardly be a 'girdle'! It seems to have arisen as a corruption of *Merak* (Beta Ursae Majoris), meaning 'loin', and to have been transferred to its new location in error. Alternatively the corruption may have occurred the other way round, because Mizar could actually depict a 'girdle' where Merak is, on the underbelly of the bear. Again, if the body of the bear is taken to represent a coffin, as it did in the eyes of some Arabian astronomers, the 'girdle' sense might be easier to reconcile. As such, the star lies roughly halfway between *Benetnasch* ('the chief of the mourners') and the near end of the bier. Star names in this part of the constellation are of uncertain meaning or provenance: compare *Alcor*, which virtually coincides with Mizar, and *Alioth*, halfway between it and *Megrez*.

Moisture, Sea of see **Humorum, Mare**

Monoceros (constellation)
The Greek name is translated in English as The Unicorn (itself a word of Latin origin). The name itself first achieved prominence from its inclusion in the star charts of 1624 compiled by the German astronomer Jakob Bartsch, although it is also found in some sixteenth-century works. Presumably the name alludes not simply to the outline of the stars (which can just about suggest a unicorn), but to the proximity of the constellations of 'hunter and hunted', that is, of *Orion, Taurus, Canis Major, Canis Minor* and *Lepus*.

Moon, The (satellite of *Earth*)
The name is so familiar that we usually give it little thought. It has a direct link with 'month', and ultimately with a root 'me-' seen in 'measure', because the Moon is the celestial body by which we measure the length of a month, after it has completed its four 'quarters' or phases, that is, when it has completed one revolution round the Earth. Words for the Moon in other languages also relate to this root, such as Greek *mēnē*, German *Mond* and Old Russian *mesyats* (which in modern Russian means 'month'). For 'month', too, links can be directly seen in Latin *mensis*, Greek *mēn* and Irish *mí*. However, there is also the 'lunar' group, typically represented by words for 'Moon' such as Latin *luna* and French *lune* (and Russian *luna*). The root here, now not much more than the initial letter 'l-', is related to 'light' and so to Latin *lux*, 'light', Greek *leukos*, 'white', Russian *luch*, 'beam', and so on. The Moon, therefore, can be regarded as either a 'measurer' or an 'illuminator'.

Morning Star, The see **Venus**

Mortis, Lacus ('lake' on Moon)
The Latin name means Lake of Death, and may have been

suggested to its namer, the Italian astronomer Francesco Grimaldi, by its proximity to the *Lacus Somniorum*, or 'Lake of Dreams', death and sleep being poetically associated. But he also may have given the name purely as a contrast to the Terra Vitae, or 'Land of Life', which no longer exists on lunar maps. Compare the name of the *Mare Frigoris* for a similar contrast. Both names date from the seventeenth century.

Moscoviense, Mare ('sea' on Moon)
The Latin name, somewhat unexpectedly, turns out to mean Sea of Moscow. But the reason for this becomes clear when it is known that the *mare* is on the far side of the Moon and was first photographed in 1959 by the unmanned Soviet spacecraft Luna 3. The name was given in honour of the Russian capital, but the type of name followed in the established pattern, even being called a 'sea', although the Moon by then had long been known to be waterless. The other famous modern *mare* name on the far side of the Moon is the *Mare Ingenii*.

Moscow, Sea of see **Moscoviense, Mare**

Musca (constellation)
The Latin name translates as The Fly, and is a subsequent alteration to the original name of Apis, The Bee, given to the constellation here in 1603 by the German astronomer Johann Bayer. It is unclear who the renamer actually was. At one time the full name of the constellation was Musca Australis, or Musca Indica, 'Southern Fly' or 'Indian Fly', because it is located in the southern hemisphere and needed to be contrasted with a former constellation, now abandoned, known as the 'Musca Borealis', or 'Northern Fly'. The latter was a star grouping in part of *Aries* and, like its southern counterpart, was also known as Apis, as well as Vespa, 'The Wasp'. It featured sporadically in star charts and atlases down to the eighteenth century, but then gradually ceased to be referred to and was no longer officially recognised. It was formed by the four stars numbered 41, 33, 35 and 39 in Flamsteed's catalogue, with these now 'riding' on the back of The Ram that is Aries.

Naos (star Zeta Puppis)
Puppis is The Stern, and part of the former much larger constellation of *Argo*. Therefore, as often happens, it is no surprise to find a star name that refers to the constellation as a whole. Here, Naos is simply the Greek word for 'ship', and the star is the brightest in its more recent constellation of Puppis. The Greek name exactly

equates to the Arabic name of Argo, which was Al Safīnah, 'the ship'.

Nashira (star Gamma Capricorni)
The name is a corruption of Arabic Al Sa'd al Nashirah, 'the fortunate one', 'the bringer of good news', which was the joint name for this star together with its near neighbour Delta Capricorni (*Deneb Algiedi*). Compare the similarly 'lucky' name of Beta Capricorni, *Dabih. Capricornus* as a whole has long been a luck-bringing constellation to astrologers, and an English text of 1386 has the words, 'Whoso is borne in Capcorn schal be ryche and wel lufyd', and much more recently the nineteenth-century English astronomer William Henry Smyth wrote that Capricorn was 'the very pet of all constellations with astrologers'. The Arabs clearly agreed!

Nata (asteroid 1086)
The asteroid was discovered in 1927 by the Russian astronomer S. I. Belyavsky, and was named after the famous Soviet woman parachutist Natalya Babushkina, killed when attempting a record-breaking jump. He could not use the names Natalya or Natasha as these had already been given to earlier asteroidal discoveries, respectively numbered 448 (at the end of the nineteenth century) and 1121 (in the 1920s). Two other women parachutists who perished were Lyuba Berlin and Tamara Ivanova, and Belyavsky similarly named asteroids after them, Lyuba (1062) and Tamariva (1084). (He could not name the latter Tamara because this name was already in use for asteroid 326, given by the Austrian astronomer Johann Palisa in the 1890s after the famous twelfth-century Queen of Georgia.)

Nebularum, Palus (feature on Moon)
The name means Marsh of Clouds, and was given to a section of the *Mare Imbrium*, so that it accords with this 'watery' name. The name first appeared on Riccioli's seventeenth-century map, and although it cannot always be found on modern maps and charts, it is still in use among astronomers.

Nectar, Sea of see **Nectaris, Mare**

Nectaris, Mare ('sea' on Moon)
The Mare Nectaris, or Sea of Nectar, can be found leading off the *Mare Tranquillitatis*, next to the *Mare Fecunditatis*. All three are 'propitious' names, appearing on Riccioli's map of 1651 (where

they lay close to the former 'Terra Sanitatis', or 'Land of Health', an equally favourable name).

Nekkar (star Beta Boötis)
One would expect one of the prominent stars in *Boötes*, The Herdsman, to have an alternative name denoting the constellation as a whole, and here it is. Nekkar is in fact a corruption of Arabic Al Baḳḳār, 'the herdsman'. It is not the brightest star in the constellation, however, because it is easily outshone by *Arcturus*, among others (despite its ranking as 'Beta', implying second brightest).

Neptune (planet)
Neptune is the eighth planet out from the Sun, and one of the most recent to be discovered. Its existence and position were calculated by the English mathematician John Couch Adams in 1845, and the following year his calculations were independently repeated by the French astronomer Urbain Le Verrier. That same year the planet was discovered at the observatory in Berlin by Johann Galle and Heinrich D'Arrest. The name of Neptune had previously been considered for the planet *Uranus* (which see for the history). Now, it was proposed for the newly discovered planet by Le Verrier. From the point of view of mythological aptness, the name could hardly have been better, given the fact that certain names had already been allocated. The genealogical line represented by the three planets *Jupiter*, *Saturn* and *Uranus* could not be continued similarly, as Uranus had no father. (In Greek mythology, Cronos, corresponding to Saturn, was the son of Uranus, and Zeus, corresponding to Jupiter, was the son of Cronos.) Therefore a name had to be sought among the brothers of Jupiter (Zeus). So it came about that Neptune, whose Greek equivalent was Poseidon, the god of the sea, entered the nomenclature of the solar system, as Poseidon was also a son of Cronos. The already established tradition of using Roman names for the gods and their planets, too, meant that it was Neptune who was selected, not Poseidon. Moreover, the planet Neptune is seen as bluish-green, the colour of the sea!

Nereid (satellite of *Neptune*)
Nereid is one of Neptune's two satellites, and was discovered long after its fellow, in 1949. The discoverer was the Dutch astronomer Gerard Peter Kuiper, who also discovered *Miranda* (which see). Clearly it required a name that somehow linked it, if possible, to both its planet and fellow satellite. Both Neptune and *Triton* have

'sea' names, and Nereid, although not an individual name, was the term used for one of the sea nymphs who were the daughters of Nereus. The association is rather more precise, however, because Amphitrite, the wife of Poseidon (Neptune), was herself a Nereid. The name can thus be regarded in addition as a 'family' one, especially as Triton was also a son of Amphitrite and Poseidon. (But why did they not simply call the newly discovered satellite Amphitrite? That would have been even better!)

Nihal (star Beta Leporis)
The name of this star is actually the same as the Arabic collective name for the four stars Alpha, Beta, Gamma and Delta Leporis, which was Al Nihāl. This meant 'the camels slaking their thirst', and referred to the neighbouring constellation of *Eridanus*, The River, in which the stars could be said to be 'drinking'. It may seem incongruous for camels to be in a constellation named *Lepus*, The Hare, but the Arabs were here thinking of a much more familiar animal, one that regularly needed to slake its thirst. The Arabs did recognise the other animal, however. See both *Lepus* and its star *Arneb*, Alpha Leporis.

Norma (constellation)
The name translates as The Level, this being a surveying instrument. Those familiar with his other names will recognise the hand of Nicolas Louis de Lacaille here, the French astronomer who introduced it in the mid-eighteenth century. He devised it from stars that were formerly in the constellations of *Ara* and *Lupus*. The star that was originally Alpha Normae has now become a member of neighbouring *Scorpius*. The original full name of Norma was Norma et Regula, The Level and Square.

North America Nebula, The (in *Cygnus*)
The name relates not to its country of discovery, but to its general outline, which does suggest the coastline of the North American continent when seen in photographs.

Northern Coalsack, The (feature in *Cygnus*)
The constellation of Cygnus lies in a section of the *Milky Way* which on dark nights shows up as having a 'split' down it. This dark lane of dust is called either the Cygnus Rift or the Northern Coalsack. The first half of this latter name serves to distinguish it from the *Coalsack Nebula* in the southern hemisphere, while 'Coalsack' fancifully indicates its dark appearance.

Northern Crown, The see **Corona Borealis**

Northern Lights, The see **Aurora Borealis**

North Star, The see **Polaris**

Nubium, Mare ('sea' on Moon)
The name translates as Sea of Clouds, and the *mare* (see Glossary) extends southwards from other 'seas' with 'watery' names, especially the *Mare Imbrium* and *Mare Vaporum*. Nor is it too far from the similarly named *Oceanus Procellarum*. This last name means Ocean of Storms, and on Riccioli's original map of 1651, on which the Mare Nubium appeared, there were other 'stormy' names that have now been abandoned, among them 'Peninsula Fulgurum' ('Peninsula of Lightnings'), 'Peninsula Fulminum' ('Peninsula of Thunders') and 'Insula Ventorum' ('Island of Winds'), all fairly closely grouped together.

Nysa (asteroid 44)
Nysa was discovered by the German astronomer Hermann Goldschmidt in 1857, the same year that he discovered *Doris* and the year before he also first detected *Europa*. The stage in asteroid-naming had now progressed from the names of mythological gods and goddesses to geographical names, although retaining the mythological association. Nysa was the name of the mountain on which Dionysus (otherwise Bacchus, the god of wine) was reared by nymphs. It has been variously located in countries as far apart as Ethiopia and India, and it may have been entirely mythical, with no actual earthly counterpart. It could even have been invented to explain the name of Dionysus, which could then be understood as 'god of Nysa'. But whatever its origin, the name is suitable enough for an asteroid!

Oberon (satellite of *Uranus*)
All but two of the names of Uranus's five original satellites come from characters in Shakespeare's plays. Oberon, discovered by Herschel in 1787, is named after the king of the fairies in *A Midsummer Night's Dream*, in which he is the husband of *Titania*, discovered that same year by Herschel. These were the first two satellites of Uranus to be discovered and named, and their names are a departure from the well-established mythological principle. Two factors seem to have accounted for this. First, all the best-known names of classical mythology had by this time been used up and allocated to other celestial bodies. Second, Herschel appears

to have lacked the interest in (or knowledge of) mythology that his earlier fellow astronomers had. He therefore turned to another literary area that he did admire and know, and this was Shakespeare. In a sense, of course, the names *are* mythological, but in a new field. In choosing the names, Herschel thus opened up a fresh naming potential, and he led the way for the Shakespearean names that were subsequently given to many of Uranus's other satellites, including *Miranda*. See also *Ariel* and *Umbriel*, however, for a slight deviation from the new norm.

Octans (constellation)
The remote and obscure constellation (centring on the celestial south pole) has an equally remote and obscure name, translating as The Octant. This was an angle-measuring instrument that was a forerunner of the better-known sextant. (It had an arc of an eighth of a circle, as compared to the sixth of a circle covered by the sextant.) However, the octant certainly had an astronomical application, for measuring the positions of stars, and in its final form was invented by the English mathematician John Hadley, who also did much to perfect the reflecting telescope. The actual name was introduced for the constellation in the 1750s by the French astronomer Nicolas Louis de Lacaille. Despite its unoriginality, we must assume that he really intended the name to be a compliment to Hadley. (How much more satisfactory it would have been, though, if he had named the group directly after him. But that would have been a break with classical tradition.) Compare the name of *Sextans*, which was not created by Lacaille.

Octant, The see **Octans**

Omega Nebula (in *Sagittarius*)
The nebula, officially numbered 17 in Messier's catalogue, is also known as the Horseshoe Nebula. Both nicknames refer to its arch-like outline, when observed through fairly powerful telescopes. This outline suggests either a Greek letter omega (see p. 19) or, in a more homely way, a horseshoe.

Oort Cloud ('cloud' of comets)
The name is given to the 'cloud' of gas, dust and comets that, according to the theory propounded by the Dutch astronomer J. H. Oort in 1950, orbit the Sun at a distance of about 40,000 astronomical units (that is, 40,000 times the distance from the Earth to the Sun, which is approximately 93 million miles).

Ophiuchids (meteor shower)
As with other meteor showers, the name refers to the constellation from which the meteors appear to 'rain' periodically, in this case *Ophiuchus* (which see). The Ophiuchids are a minor shower, with 'peak performance time' in the second half of June annually.

Ophiuchus (constellation)
The name is an ancient one, dating from classical times, and is usually rendered in English as The Serpent Holder. The name is a Greek one (in the original, *Ophioukhos*, from *ophis*, 'snake' or 'serpent', and *ekho*, 'I hold'), and its Latin equivalent was Serpentarius. The outline depicts a man encoiled by a serpent, the latter represented by the two halves of the adjacent constellation of *Serpens*, with its head one side and its tail the other. Almost certainly, the Greek and Roman astronomers identified the figure with that of Aesculapius (Asclepius), the god of healing, who was the ship's doctor on the *Argo*. On the other hand, the figure may have been held to have some connection with the important neighbouring constellation of *Hercules*, perhaps in connection with one of his Twelve Labours. A serpent-holder or serpent-tamer is, after all, something of a hero or 'strong man', and in his Second Labour Hercules did kill the Water Snake, or *Hydra*, although astrographically the latter constellation is light years away from the others mentioned above, and is in the southern hemisphere. The equivalent Arabic name was Al Ḥawwāʿ, 'the snake-charmer', a corruption of which occurs as the second word in the name of the constellation's most prominent star, *Ras Alhague* (Alpha Ophiuchi). The most famous star in Ophiuchus, however, is not this but *Barnard's Star*.

Orientale, Mare ('sea' on Moon)
The name means straightforwardly Eastern Sea, and points to the location of this particular *mare* on the extreme limb (edge) of the Moon, where it in fact extends round to the far side. However, as seen from Earth, the Mare Orientale is on the extreme western edge, not eastern, because at the time of naming, the 'east' of the Moon was regarded from the point of view of someone standing on it. (Later, the International Astronomical Union 'reversed' east and west – hence the apparent misnomer.) Despite its Latin wording, the name is actually a modern one, given to the *mare* in 1946 by the two English astronomers who discovered it, H. P. Wilkins and Patrick Moore. The name is a 'compass' one on the same basis as that of the older *Mare Australe*.

Orion (constellation)

A noble constellation with a noble name, The Hunter, containing some of the finest and brightest stars in the sky, including *Betelgeuse* (Alpha Orionis) and *Rigel* (Beta Orionis), to say nothing of the Orion Nebula, centring on Theta Orionis, known as the *Trapezium*. In classical mythology, Orion was the giant hunter known for his handsomeness and his many adventures. His stories seem to have been invented to account for the position and apparent motion of the stars in his own and neighbouring constellations. Like all hunters, he is accompanied by his dog (*Canis Major*), while a hare (*Lepus*) flees before him. But if a hare was too puny an animal to be chased by the great Orion, then doubtless it was the bull (*Taurus*) that he was hunting. One story, too, tells how he was stung by a scorpion (*Scorpius*) and died from its sting, and that Orion was then placed in the sky so that he would set as Scorpius rises. Again, according to some, Orion pursued not only wild game but also the daughters of Atlas, the *Pleiades* as well. And there he is still following them in the adjacent constellation of Taurus. Surrounded by all these animals and people, Orion is himself a glittering personage, with his belt formed of three bright stars, respectively Zeta, Epsilon and Delta Orionis, otherwise *Alnitak*, *Alnilam* and *Mintaka*. In his right hand he holds a sword, although the star named for it, *Saiph*, has somehow slipped down to his foot, while his left hand holds a shield, composed of a curve of six stars all lettered as Pi Orionis, with which he wards off the snorting Taurus. The Arab astronomers knew the constellation simply as Al Jabbār, 'the giant', and this name also appeared in the writings of some Greek and Roman authors as Gigas. Similarly the Arabic name for the three stars of Orion's Belt was Al Nijād, 'the belt', or simply Al Nisak, 'the line'. Other peoples and races have distinguished the three by quite different names, such as (in translated form) Three Deer (North American Indians), Three Zebras (Hottentots), Three Men (Eskimos), Three Sisters (White Russians) and Three Ploughmen (some parts of Germany). The three have understandably also been seen by some races as the Three Wise Men. Orion and his components seem to have acquired a range of names as varied as the stars and other celestial objects that comprise him.

Orionids (meteor shower)

The meteors are named after the constellation of *Orion* from which they radiate annually, in the second half of October, with their actual 'sparkoff point' being close to the border with *Gemini*.

Owl Nebula (in *Ursa Major*)
Catalogued as M 97, the nebula contains two stars that do fancifully suggest the face of an owl when photographed with a powerful instrument.

Painter's Easel, The see **Pictor**

Pallas (asteroid 2)
Pallas was the second asteroid to be discovered, and its discoverer and namer in 1802 was the German doctor and astronomer Heinrich Olbers. Following the precedent set by the name of *Ceres*, discovered the previous year, Olbers chose the name of a goddess in classical mythology. Whereas Ceres, however, was the main name of a Roman goddess, Pallas was the byname or 'title name' of a Greek goddess, Athene. Why did he not actually use the name Athene? It seems likely, at this early stage of asteroid-naming, that Olbers felt the name could be misunderstood as being that of Athens, so therefore chose the goddess's secondary name. (In German, too, the names of city and goddess are closer than they are in English, respectively Athen and Athene.)

Parthenope (asteroid 11)
Parthenope was discovered in 1850 by the Italian astronomer Annibale de Gasparis, who the previous year had discovered *Hygeia* and who would go on to discover, among others, asteroids 13, 15, 16 and 24 (respectively Egeria, Eunomia, Psyche and *Themis*). The name preserves the mythological pattern already established for the first ten asteroids, as Parthenope was one of the sirens. (There was also a Parthenope who bore a son to Heracles.) But the name had added point for de Gasparis, because he lived in Naples, and made his discoveries there, and Parthenope was the ancient original name of Naples. A practical commemorative note was thus now being introduced in the new names devised for asteroids!

Pasiphaë (satellite of *Jupiter*)
Pasiphaë is the seventh named satellite of Jupiter, travelling outwards, and it was discovered in 1908 by the Belgian astronomer P. J. Melotte, at Greenwich Observatory. As with the names of Jupiter's other satellites, Pasiphaë was a mythological character connected with Zeus (Jupiter). In her case, she was the daughter of Zeus, at least according to one account. (This Pasiphaë was thus not the better-known one who coupled with Minos to produce the Minotaur, and it seems strange that a confusing name from a subsidiary legend should have been chosen for the discovery. But

equally, the namers will have needed a name ending in '-e', for as explained in the Introduction, p. 22, the names of minor Jovian satellites that have a retrograde motion, as Pasiphaë does, all end in this particular vowel, while the others end in '-a'.)

Patientia (asteroid 451)
The asteroid was discovered in 1899 by the French astronomer Auguste Charlois, and was given one of a number of Latin abstract 'moral' or 'virtue' names used for several asteroids. This one, fairly obviously, means 'patience'. Among others were asteroids 474 (Prudentia, 'circumspection'), 109 (Felicitas, 'happiness'), 306 (Unitas, 'unity'), 490 (Veritas, 'truth'), 902 (Probitas, 'honesty') and 996 (Hilaritas, 'cheerfulness'). There was even a basic Virtus (494), but this was probably taken to mean 'manliness' rather than just 'virtue'.

Patroclus (asteroid 617)
Patroclus was discovered in 1906 by the same astronomer who the following year was to locate *Hector*. This was the German August Kopff. It was argued that there should be more than one *Trojan* asteroid, similar to *Achilles*, and so moving in the orbit of *Jupiter*. After a feverish search, one such asteroid was located. So what better name to give it than Patrocles, as in the legendary Trojan War, Patrocles was Achilles' best friend. (Alas, Patrocles was to be killed by Hector. But the names of all three warriors are now assured in the sky.)

Pavo (constellation)
The Latin name translates as The Peacock, and it was introduced for this constellation by the German astronomer Johann Bayer in 1603. As a celestial bird, it is accompanied in this region of the sky by *Apus*, *Grus*, *Phoenix* and *Tucana*. In Greek mythology, the peacock was sacred to Hera, the Greek goddess from whose breast the *Milky Way* sprang. The constellation has no named stars.

Peacock, The see **Pavo**

Pegasus (constellation)
The ancient name is that of the famous winged horse in classical mythology, and the constellation was alternatively called The Winged Horse in English until relatively recently. The Greeks also called it simply Hippos, 'the horse', while the Roman equivalent for this was Equus. The main Arabic name for the stars here, however, was Al Dalw, 'the water bucket', referring not to all the

stars in the constellation but to those four that form the distinctive square, with Alpha, Beta, Gamma and Delta Pegasi marking its four corners. (The respective names for these are *Markab*, *Scheat*, *Algenib* and *Sirrah*. This last, however, is now no longer Delta Pegasi, but Alpha Andromedae, as it is just over the border in *Andromeda*. The first three names relate directly to the figure of a horse, or a part of it, as does the name of Epsilon Pegasi, *Enif*. If it were not for these, a casual observer might find it difficult to detect the outline of a horse here, winged or not, for it is hardly obvious.)

Perseids (meteor shower)
As their name indicates, the meteors appear annually to radiate from a point in the constellation of *Perseus* (which see). They are usually best seen on or near 12 August.

Perseus (constellation)
This is a much more vivid depiction than that of *Pegasus*. Perseus was the famous hero of Greek mythology who rescued *Andromeda* from the sea monster, *Cetus*, and who earlier had slain the Gorgon Medusa. As he is represented in the night sky, he can be seen holding the head of Medusa in one hand, with her evil eye marked by the winking star *Algol*. The name is thus an ancient one, dating from classical times. The Arabian astronomers either adopted the Greek and Roman name (mostly in the form Fersaus, as Arabic has no letter 'p') or else called the constellation Hāmil Rā's al Ghūl, 'the bearer of the demon's head', with the last part of this giving the name of Algol. Astrologers who preferred biblical names have correspondingly sometimes called it David with the Head of Goliath. Despite the close connection of Andromeda and Cetus with Perseus in the stories about him, these two constellations do not border his, although they are both in the northern hemisphere, as he is.

Phecda (star Gamma Ursae Majoris)
Like some other stars in the constellation of *Ursa Major*, Phecda refers to a part of the body of The Great Bear. It marks the top of his rear leg, and derives from Arabic Al Faḥdh, 'the thigh'. Compare the names of *Merak* and *Megrez*, two fairly close companions of Phecda.

Pherkad (star Gamma Ursae Minoris)
The name is Arabic in origin, deriving from Al Farḳadain, 'the two calves'. The Arabs regarded this star and the nearby one we now

know as 11 Ursae Minoris as a related pair, and for many years the English names for the two were Pherkad Major and Pherkad Minor. We now know that the second star is quite unrelated to the first, however.

Phobos (satellite of *Mars*)
Mars has two satellites (the other is *Deimos*), both of which were discovered by the American astronomer Asaph Hall in 1877. He named them respectively after the mythical characters who drove the war chariot of Mars. These were Phobos (fear) and Deimos (terror), the names coming straight from the Greek. The names were not chosen simply for their 'threat' value, but because they were actually those of the sons of Ares (Mars)!

Phoebe (satellite of *Saturn*)
Phoebe, the outermost of Saturn's named satellites, was discovered in 1898 by means of photography (the first time this had been done) by the American astronomer William Henry Pickering. The name was carefully chosen to accord with those of earlier satellite discoveries for this planet, because Phoebe, in classical mythology, was not only the daughter of a *Titan*, but also the goddess of light. Most of Saturn's satellites have the names of Titans, and a name associated with light is obviously suitable for a shining celestial object. (Compare *Hyperion*, another of Saturn's satellites, in this respect.)

Phoenicids (meteor shower)
The name indicates that the meteors appear from the constellation of *Phoenix* (which see), which they do annually, albeit as a minor shower, in early December.

Phoenix (constellation)
The constellation of the southern skies has a name with an obvious meaning (The Phoenix) introduced by the German astronomer Johann Bayer in 1603. The Arabs knew the stars here as Al Zaurak, 'the boat', seeing the formation as representing a boat tied up alongside the neighbouring constellation of *Eridanus*, The River. Bayer may have devised the name simply for its exotic attraction (compare his birds of paradise, crane, peacock and toucan, also introduced by him, as *Apus*, *Grus*, *Pavo* and *Tucana*). At the same time, the name is astrologically successful, because the phoenix was the fabulous Egyptian bird associated with a regular cycle of years (usually 500), at the end of which it burnt itself and its nest to ashes only to arise from the fire with new life to commence a new cycle.

The Egyptians, in fact, regarded the bird as a symbol of immortality, and depicted it on their coins. Moreover, the actual outline of the stars in the constellation do quite readily suggest the image of a bird rising into the air with its wings outstretched, even though no star is now individually named.

Piazzia (asteroid 1000)

An asteroid with this catalogue number deserves a special name! It has got it, for the name represents that of the Italian astronomer Giuseppe Piazzi, who had discovered the first asteroid, *Ceres*, in 1801. Piazzia, with its distinctive 'feminine' ending ('-ia'), was itself discovered in the 1920s, when there was a vogue for naming asteroids after astronomers. For example asteroid 1001 was called *Gaussia*, after the German Karl Gauss, and 1002 was named Olbersia, after Heinrich Olbers, who had himself discovered *Pallas* and *Vesta* soon after Piazzi had made his pioneering move.

Pictor (constellation)

The Latin name was introduced for the constellation in the mid-eighteenth century by the French astronomer Nicolas Louis de Lacaille, originally in the full form of Equuleus Pictor. This last provides the standard English name for it today, The Painter's Easel. The three brightest stars in the faint constellation, Alpha, Beta and Gamma Pictoris, do perhaps suggest the shape of an easel. The full name of the constellation must not be confused with that of *Equuleus*. The only link between the two names is a linguistic one: both Latin names are related to *equus*, 'horse', as English 'easel' is related to 'ass'. The horse and the ass carry a burden just as the easel carries a picture (or the canvas for one).

Pisces (constellation)

The name is an ancient one, meaning The Fishes, and the constellation represents a pair of fishes tied together by their tails, with the 'knot' marked by the star *Al Rischa* (Alpha Piscium). The depiction could have links with more than one legend in classical mythology. Perhaps it is the one in which Aphrodite and her son Eros (known to the Romans as Venus and Cupid) leapt into the river Euphrates and turned into fishes to escape the monster Typhon. The present name, as of many constellations, is the Latin one. The Greek name for it was either Ichthye or Ichthyes. (The first of these is grammatically dual, that is, denoting two, and the second plural.) The Arabic name was similar, either Al Samakatain (dual) or Al Samakah (plural), both translating as 'the fishes'. As already pointed out for *Aquarius*, this is a 'watery' part of the skies, and other constellations nearby include *Capricornus* and *Cetus*. Compare also *Piscis Australis*.

Piscis Australis (constellation)
The name, which is sometimes given as Piscis Austrinus, means The Southern Fish. Like *Pisces*, it is an ancient one, and indeed astrologically the fish represented here has been seen as the parent of the zodiacal one, Pisces. It was usually thought of in classical times as drinking the water that flowed from the urn of nearby *Aquarius*. It is 'Southern' not necessarily to distinguish it from the more northerly Pisces, but possibly because it was regarded as representing a particular fish here, even a sea creature that has already given its name to its own constellation, such as *Cetus*, *Delphinus*, *Hydra* or even *Scorpius*. The Greeks called it either simply Ichthys, 'the fish', or similarly added an adjective, such as Ichthys megas, 'big fish', or, like the Romans, Ichthys notion, 'southern fish'. The Arabs adopted the Greek name and called the constellation Al Ḥūt al Janūbiyy, 'the large southern fish'. (For the first part of this, compare the name of Alpha Piscis Australis, *Fomalhaut*.)

Planet X (planet yet to be discovered)
Today the name is popularly used for any planet that may still lie undiscovered in the Solar System outside the orbit of the outermost known and named planet, *Pluto*. The existence of such a planet is perfectly possible. Pluto was itself a 'Planet X' for Percival Lowell, who, alas, died before it was discovered in 1930. 'X', of course, is a symbol for something unknown or that has to be discovered or determined. If a tenth planet *is* discovered, the selection of a suitable name for it will clearly be very important. It will probably be mythological (at one stage 'Minerva' was a candidate.)

Plaskett's Star (in *Monoceros*)
The star is a binary (see Glossary) and one of the most famous in the constellation. It is named after the Canadian astronomer, John S. Plaskett, who discovered in 1922 that it is the most massive stellar system yet known, with the mass of each component estimated as being over fifty-five times that of the Sun.

Pleiades (star cluster in *Taurus*)
Known popularly as the Seven Sisters, the Pleiades are the most famous and the brightest star cluster in the night sky. The stars are individually named after the seven nymphs who were the daughters

of Atlas and Pleione in classical mythology, with their mother given the overall name of the group. They are now regarded as actually being nine in number, because their parents are included. The names of the daughters are thus *Alcyone* (the brightest), Celaeno, Electra, Taygeta, Maia, Asterope and Merope. There are two main accounts as to how the sisters found their way to the sky. The first says that Zeus put them there because they had died from grief over the death of their half-sisters, the *Hyades* (also in Taurus). The second says that Zeus rescued them from the lusty *Orion*, who was pursuing them and their mother. Pictorially the second version makes the better dramatic (or melodramatic) sense, because Orion is still there in the sky chasing them! The Greeks also associated the name with the verb *plein*, to 'sail', because the Pleiades rose in the sky annually in May, when the sailing season began. The two main Arabic names for the group were either Al Thurayyā, 'the chandelier', or simply Al Najm, 'the constellation' (that is, '*the* constellation'). Some modern astronomers, however, associate the first name with that of Hathor, the Egyptian goddess of the sky. A popular name in many countries for the stars has been Hen with her Chickens (in the appropriate language), and other varied names have been recorded, from the Australians' Young Girls (playing to dancing young men) to the South Africans' Hoeing Stars (indicating a time to hoe, like the sailing time of the Greeks). The magic number seven, too, has produced many sets of varied individual names for the stars, from the Seven Sages to the seven American poets of the War of Independence, members of the so-called 'Connecticut Wits'!

Plough, The see **Ursa Major**

Pluto (planet)

Pluto was discovered only in 1930, and so the naming of the planet is one of the most important events this century in astronymical terms. Its discoverer was the American astronomer Clyde Tombaugh, but the calculations for the existence of Pluto had been made many years earlier by his compatriot, Percival Lowell, who called it his '*Planet X*'. Alas, Lowell did not live to see the realisation of his computations. (Alas, too, Lowell is all too frequently remembered today for his firm belief in Martians, and their construction of canals on *Mars*.) Once the existence of the ninth planet in the Solar System had been confirmed, the next important matter was how to name it. It was clear that Lowell had played a crucial role in the eventual discovery, and his name was one of those proposed for the planet. However, this would have broken the well established principle of related mythological names. How to combine both? To honour Lowell, it was decided to call the planet by the name of a mythological character that began with his initials, P.L. The discovery itself was made on 18 February 1930, when Tombaugh was examining photographic plates he had taken on 21, 23 and 29 January. The actual announcement of the discovery was not made until 13 March, however, the exact anniversary of Lowell's birth date in 1855. (He died in 1916.) Among the many names proposed for the planet (another, apart from 'Lowell', was 'Minerva', but this had already been allocated to an asteroid), 'Pluto' was the winner. Not only did it begin with the required two letters but, in mythology, Pluto was the god of the underworld who was the brother of Poseidon (*Neptune*) and of Zeus (*Jupiter*), and these two names already existed for planets. The name breaks with the tradition of giving the Roman names of gods, equivalent to the Greek characters, simply because Pluto had no Roman equivalent. But this does not matter: the name was also apt because as that of the god of the underworld, it could now be used for the planet that was the outermost (as far as was known) in the Solar System, and so was the most remote and in 'eternal night'. The actual name is on record as having been suggested by an 11-year-old Oxford schoolgirl, Venetia Burney. She proposed the name to her grandfather over breakfast on 15 March, two days after the announcement of the discovery. The name was approved, and the following month the President of the Astronomical Society offered 'congratulations to the suggester of the name Pluto, now adopted'. (It was a brother of Venetia Burney's grandfather who had proposed the names of *Deimos* and *Phobos* for the two satellites of *Mars*.) For more details of the discovery of Pluto, and of rejected names, see either Clyde Tombaugh's *Out of the Darkness: The Planet Pluto*, or Tony Simon's book (written for children, but informative enough to be of interest to adults) *The Search for Planet X* (see Bibliography for both titles).

Pointers, The see **Ursa Major**

Polaris (star Alpha Ursae Minoris)
The constellation of *Ursa Minor* contains the (celestial) north pole, at least at present, and the star Polaris lies very conveniently within 1° of this. Its Latin name, obviously enough, means 'polar', and in English it is equally well known as simply the Pole Star. In later scientific work, the star is also called Cynosura, from the Greek

name (literally translating as 'dog's tail') for the constellation as a whole. The early Greek name for Polaris, however, was Phoenice, again taking the name of the entire constellation. The Arabs knew the star by several names, of which the most popular was Al Ḳiblah. This means 'direction towards which one turns in prayer' (towards Mecca), because the Arabs used the star as an orientation point when they were out of their native surroundings and had to turn to Mecca when praying, as all Muslims do. (At this period, Polaris would have been about 5° away from the north pole.) They also called it Al Ḳuṭb al Shamāliyy, literally 'the axle of the north', because they regarded the heavens as revolving round the north pole. But more simply, they knew it equally as Al Kaukab al Shamāliyy, 'the star of the north'. (Compare *Kochab*, Beta Ursae Minoris.) Precession (see Glossary) will bring Polaris closest to the north pole in about the year 2100, after which it will start moving away again. For more details regarding its alternative names, see *Ursa Minor*.

Pole Star, The see **Polaris**

Pollux (star Beta Geminorum)
The name will be familiar as the 'other half' of The Twins, the first being *Castor*, who are represented as the constellation of *Gemini*. Pollux is actually brighter than Castor, and one would have expected the names to have been assigned the other way round. It was the German astronomer Johann Bayer, however, who gave the stars their Greek designations in the early seventeenth century, calling Castor Alpha and Pollux Beta, so he is really to blame for not labelling Alpha Geminorum as the brightest star in the constellation.

Porrima (star Gamma Virginis)
Unusually, several named stars in the constellation of *Virgo* are Latin in origin, not Greek, and not even Arabic, as is usually the case. Here, Porrima was the name of a goddess of prophecy, one of the attendants of the prophetess Carmente when she came from Arcadia. (The story can be found in Ovid's *Fastes*.) Porrima, therefore is the 'lucky star' of the group. Possibly the 'luck' was specially associated with the harvest, as Virgo herself was. *Spica* (Alpha Virginis) and *Vindemiatrix* (Epsilon Virginis) also have harvest associations.

Praesepe (star cluster in *Cancer*)
Officially catalogued as NGC 2628, or numbered 44 in Messier's

listing, Praesepe has a Latin name that means 'crib' or 'manger'. The name does not refer to the appearance of the cluster, or its outline, but was doubtless devised to provide a pictorial 'home' for the two stars above and below it called respectively *Asellus Borealis* (Gamma Cancri) and *Asellus Australis* (Delta Cancri), otherwise 'northern donkey' and 'southern donkey'. The Greek name for the cluster, Phatne, had an identical meaning, as did the Arabic, translated from the Greek, which was Al Ma'laf, 'the stall'. The cluster is also known in English by the nickname of The Beehive. Here the name more closely suggests the formation, even though the cluster appears to the naked eye simply as a blurred patch. This could, however, suggest bees swarming round a hive.

Procellarum, Oceanus ('ocean' on Moon)
The name means Ocean of Storms, and is a suitably 'watery' name to accompany the various regions into which it leads, such as the *Mare Imbrium*, *Mare Nubium* and *Sinus Roris*. It is an 'ocean' rather than a 'sea' simply because of its vast size. It lies in the western half of the near side of the Moon. The name is on Riccioli's map of 1651.

Procyon (star Alpha Canis Minoris)
The star has a Greek name, meaning literally 'before the dog'. This refers to the fact that the constellation of *Canis Minor* rises before that of *Canis Major*, and the name itself thus was also used, as often happened of the most prominent star, for the whole constellation. The Romans had a name for the star that was exactly the same in meaning, Antecanis, while the Arabs knew it as Al Shi'rā al Shāmiyyah, 'the bright star of Syria' (see *Canis Minor* for the explanation of this). In fact many alternative names for the constellation as a whole were used for the individual star.

Protrugeter see **Vindemiatrix**

Proxima Centauri (star in *Centaurus*)
This is the name of the star that is a close companion in the sky of *Rigil Kentaurus* (Alpha Centauri) but that is not in the same telescopic field of view. Its name means 'nearest (star) of Centaurus', referring to its unique status of being the nearest star to us beyond the Sun. It is technically a red dwarf (see Glossary), and was discovered by the Scottish astronomer Robert Thorburn Innes in 1913.

Ptolemaeus (crater on Moon)
Ptolemaeus is an extensive walled plain in the south central section

of the near side of the Moon, and is one of the many 'craters' to be named after prominent scientists and astronomers, in this case after Ptolemy, the last great astronomer of classical times. The name first appeared on the map of the Moon made by Riccioli in 1651, based on that of his pupil Grimaldi.

Puppis (constellation)
Puppis, meaning The Stern, is the largest of the four sections into which the original constellation of *Argo* was divided by the French astronomer Nicolas Louis de Lacaille in the mid-eighteenth century. (The others are *Carina*, *Pyxis* and *Vela*.) The outline of the brightest stars of the constellation do actually suggest the stern section or poop of a sailing ship.

Putredinis, Palus (feature on Moon)
The name translates as the hardly attractive Marsh of Decay. The designated region forms part of the *Mare Imbrium*, and so is one of the 'watery' names that are characteristic of the western half of the near side of the Moon. It appears on Riccioli's map of 1651, based on that of his pupil Grimaldi.

Pyxis (constellation)
Pyxis means The Compass, and is the smallest of the four sections into which the ancient constellation of *Argo* was divided for convenience in the 1750s by the French astronomer Nicolas Louis de Lacaille. (Compare *Puppis*.) The constellation is not an impressive one, but the line formed by its three brightest stars Alpha, Beta and Gamma Pyxidis could be seen as a compass needle.

Quadrans Muralis (former constellation)
The constellation, whose name translates as The Mural Quadrant, was formerly in the area of sky now occupied by the northern part of *Boötes*, and it is included here solely on account of the *Quadrantids* (which see). The constellation was actually introduced by the French astronomer Joseph Lalande in 1795, and was named after the instrument with which he and his nephew, Michel Le Français, had observed the stars in this area. A mural quadrant is a quadrant (an instrument for measuring an arc of a quarter of a circle, or 90°) fixed to a wall.

Quadrantids (meteor shower)
The name is that of an annual meteor shower radiating from that section of *Boötes* that was formerly the constellation of *Quadrans*

Muralis. The shower is at its most spectacular in the New Year, with its maximum or 'peak' on 3 January.

Rainbows, Bay of see **Iridum, Sinus**

Rains, Sea of see **Imbrium, Mare**

Ram, The see **Aries**

Ras Algethi (star Alpha Herculis)
The name for this red supergiant, one of the largest known stars in the sky, comes from Arabic Al Rās al Jāthīyy, 'the head of the kneeler', referring to its location at the head of the kneeling figure of *Hercules*. The name serves as a reminder that the kneeling figure was originally anonymous, before it came to be particularly associated with Hercules.

Ras Alhague (star Alpha Ophiuchi)
The name derives from Arabic Rās al Ḥawwāʿ, 'head of the snake charmer', because this is what it marks in the depiction of The Serpent Holder who is the constellation *Ophiuchus*.

Rastaban see **Alwaid**

Regulus (star Alpha Leonis)
The name is the only Latin one in the constellation of *Leo*, and it means 'the little king'. The star has long been regarded as the one that 'ruled' or controlled the affairs of the heavens, and although this particular name was introduced by Copernicus in the sixteenth century, it closely follows the former Latin name of *Rex*, 'the king', and Greek *Basiliscos*, 'kingly'. (The star itself has been connected by some with the ancient king, Amagalaros.) The Greek name was in turn translated, with exactly the same meaning, by the Arabs, who called it Malikiyy. In English, too, it has for some time been known as the Royal Star. It is located in the sickle-shaped line of stars that form the head of The Lion, where it is the southernmost and brightest.

Reticulum (constellation)
The name of this constellation was introduced in the mid-eighteenth century by the French astronomer Nicolas Louis de Lacaille, who wished to commemorate the instrument called a 'reticle' with which he observed the stars of the southern hemisphere. A reticle was (and still is) a set of crisscrossing lines etched or drawn on a special lens that is then inserted into a telescope, thus enabling the observer to make accurate measurements. The usual English name for the

constellation, however, is The Net. The original full name of the constellation was Reticulum Rhomboidalis, 'the rhomboidal net', referring to its lozenge shape.

Rhea (satellite of *Saturn*)
Rhea was discovered in 1672 by the Italian astronomer Giovanni Domenico Cassini, who the previous year had discovered *Iapetus* and who would go on to locate *Dione* and *Tethys*. Like these two last names, Rhea was the name of one of the *Titans* in classical mythology, and in her case she was the wife of Cronos (Saturn), who was actually her brother.

Rigel (star Beta Orionis)
The Arabic name derives from the title Rijl al Jawzah, 'leg of the Jauzah', with the latter word usually translated as 'giant' and so meaning *Orion*. The star, which is the brightest in the constellation, marks the left foot of Orion, as he stands facing us, with his right foot marked by *Saiph* (now a misnomer for it). Another Arabic name for Rigel was Ra'i al Jawzah, 'herdsman of the Jawzah', because it was seen as 'herding' four camels, respectively represented by Alpha, Gamma, Delta and Kappa Orionis, otherwise *Betelgeuse*, *Bellatrix*, *Mintaka* and *Saiph*. Compare also *Rigil Kentaurus*.

Rigil Kentaurus (star Alpha Centauri)
The name of the star derives from Arabic Al Rijl al Kentaurus, 'the foot of the centaur', because this is what it prominently represents (on the right foreleg) of The Centaur as he stands in the sky. Closely associated with Rigil Kentaurus is the star *Proxima Centauri*. The name is little used by astronomers today. For a similar name, compare *Rigel*.

Ring Nebula (in *Lyra*)
Officially catalogued as NGC 6720, and also known by its Messier number of 57, the Ring Nebula has a fairly obvious nickname referring to its appearance when seen through a telescope, when it resembles a sort of heavenly smoke ring. It is formed mainly by gases thrown off by the star in its centre.

River, The see Eridanus

Robur Carolinum (former constellation)
The name, which translates as Charles's Oak, is worth recording as a creation of the famous English astronomer Edmond Halley, better known for his comet. He introduced it in 1679 in what is

now the constellation of *Carina* (but was then *Argo*), wishing to commemorate the famous oak tree in which his royal patron, Charles II, had hidden in 1651 after his defeat by Cromwell in the Battle of Worcester. This tribute won Halley a degree of Master of Arts at Oxford (where he had been a student) by special decree of the king. The French astronomer Nicolas Louis de Lacaille, however, complained that Halley had ruined Argo by his introduction of the new constellation, and despite support for it on the part of a few fellow astronomers, such as the German Johann Bode, the name was not widely accepted and was ultimately abandoned.

Roris, Sinus ('bay' on Moon)
The name translates as Bay of Dew, and is thus one of the 'watery' names found in this western half of the near side of the Moon, where it runs southwards from the *Mare Frigoris* to the *Oceanus Procellarum*. It appears on Riccioli's map of 1651, which he had based on the one compiled by his pupil, Grimaldi.

Rosette Nebula (in *Monoceros*)
The nebula, officially catalogued as NGC 2237, is nicknamed from its rather striking appearance as a 'rosette' surrounding the star cluster NGC 2244. The name refers not only to its shape but also to its pink colour, when seen on a long-exposure photograph.

Rotanev (star Beta Delphini)
Readers who have been working more or less systematically through this book will see this name and may expect another Arabic origin. But they will have a surprise, because the name is actually the backwards form of Venator, which was the Latinised surname of the Italian astronomer Niccolo Cacciatore (meaning 'hunter')! He was the assistant to Giuseppe Piazzi (who discovered the asteroid *Ceres*), and the star was named as a tribute to him, as was Alpha Delphini, *Sualocin*. Both names first appeared in Piazzi's catalogue of 1814. And what more suitable name in honour of a man who was a 'hunter' of the stars?

Ruchbah (star Delta Cassiopiae)
This name, like that of *Schedar*, refers to a part of the figure of *Cassiopeia* as she appears sitting in her chair in the skies. It comes from Arabic Al Rukbah, 'the knee', because this is the feature of the Queen of Ethiopia that it marks. Compare the alternative name of *Al Rami*, which is Rukbat, with exactly the same sense.

Rukbat see **Al Rami**

Sadachbia (star Gamma Aquarii)
The name is Arabic in origin, and probably represents Al Sa'd al Aḣbiyah, 'the lucky one of the tents', referring to the fact that when the star appeared, in the spring, the weather would improve and nomads would be able to pitch their pastures. Some astronomers regarded the nearby stars Zeta, Eta and Pi Aquarii as representing the actual tents. For other 'lucky star' names in *Aquarius*, see *Sadalmelik* and *Sadalsuud*.

Sadalmelik (star Alpha Aquarii)
The name is a corruption of Arabic Al Sa'd al Malik, 'the lucky one of the king'. For the cause of all this 'luck' in this area of *Aquarius* see *Sadalsuud*, which has the luckiest name of all.

Sadalsuud (star Beta Aquarii)
The name represents Arabic Al Sa'd al Su'ud, 'the lucky one of the lucky'. The 'luck' here, and for *Sadachbia* and *Sadalmelik*, was not simply that they appeared when the winter was over and that the spring, with its fairer weather, had arrived. There is more of an astrological explanation than that. The fact was that it was in the region of these stars that the so-called 'stations' of the Moon occurred, those sections of the lunar zodiac where the Moon was seen to pass between the stars. When the Moon reached a particular 'station', it was considered lucky for anyone who needed luck, such as poor people, travellers or (especially with regard to these stars) the monarchy.

Sadr (star Gamma Cygni)
The constellation of *Cygnus* is The Swan, and Sadr is the star that represents the breast of this bird, from Arabic Al Sadr, 'the breast'. Compare names such as *Deneb* and *Gienah* that were given to stars marking other parts of the bird's body. Compare also *Schedar*.

Sagitta (constellation)
This name, not to be confused with that of *Sagittarius* (which see), is an ancient one, translating as The Arrow. The stars here do actually form an arrow shape, and appear to be 'flying', appropriately enough, between *Cygnus* to the north and *Aquila* to the south, respectively The Swan and The Eagle. No doubt there is more than one classical story to explain the presence of the arrow here. Perhaps Hercules shot it, during one of his Twelve Labours. One would like to think that The Archer did (see *Sagittarius*), but he is nowhere near, and appears to be quite unconnected with it.

Sagittarius (constellation)
The constellation was known in classical times, with its name translating into English as The Archer. The depiction here is that of a
centaur, a creature that was half beast, half man. However, this
centaur is quite different from the one represented in *Centaurus*.
He was kindly and wise and helpful, and carried no weapon.
Sagittarius, on the other hand, appears in a menacing attitude, and
holds a bow and arrow in his hand, aiming it at the heart of the
neighbouring *Scorpius*, The Scorpion! (His bow is marked by the
stars Lambda, Delta and Epsilon Sagittarii, otherwise respectively
Kaus Borealis, *Kaus Meridionalis* and *Kaus Australis*, with the
arrow indicated by Gamma Sagittarii, *Al Nasl*.) No doubt the idea
of an archer here goes back several thousand years, perhaps to the
Sumerians, whose archer god of war was Nergal. One of the
common Greek names for the constellation, with the same
meaning, was Toxotys. The Arabs chiefly used the name Al Ḳaus,
'the bow', regarding this part as representing the whole constellation. Earlier Arabic names for star groups here, however,
included Al Naʿām al Wārid, 'the going ostriches' (for Gamma,
Delta, Epsilon and Eta Sagittarii) and Al Naʿām al Ṣādirah, 'the
returning ostriches' (for Sigma, Zeta, Phi, Chi and Tau Sagittarii).
These groups were seen as going to and returning from the *Milky
Way*, as the celestial river, with Lambda Sagittarii thought of as
their keeper.

Sails, The see **Vela**

Saiph (star Kappa Orionis)
The original Arabic name for this star, which marks the right foot
of *Orion*, as he stands facing us, was Rijl Jawzah al Yamnā', 'the
right leg of the Jauzah' (of Orion). Somehow, however, a naming
error occurred, and Kappa Orionis came to be called Saiph, from
Al Saif, 'the sword', which is actually represented by Eta Orionis,
or at worst Iota Orionis, where this weapon hangs from Orion's
belt. See also *Rigel*, which marks his left foot.

Saturn (planet)
Saturn was the most remote planet known in ancient times. The
Greeks called it Phainon, 'the shining one', originally, before
Aristotle renamed it Cronos, after the Greek god who was the ruler
of the *Titans* and the son of *Uranus* (the sky). The Romans then
came to call the planet by the name of their god who was the
equivalent of Cronos, that is, Saturn. Cronos was a suitable name

for the planet which came next in order after *Jupiter*, because Cronos was the father of Zeus (the Greek god to whom the Roman Jupiter corresponded). The three planets Jupiter, Saturn and Uranus thus represent the three mythological generations of son, father and grandfather. (Unfortunately the line could not be taken any further, because Uranus had no father. See *Neptune* for what happened!)

Saturn Nebula (in *Aquarius*)
The nebula, catalogued as NGC 7009, is so nicknamed since it has a band across it that makes it look something like the planet *Saturn* when seen telescopically.

Scales, The see **Libra**

Sceptre and Hand of Justice (former constellation)
This was the name of the constellation in the region now occupied by *Lacerta* before it became *Frederici Honores* in 1787. The introducer of the first name was the French astronomer Augustin Royer, in 1679, with the aim of paying a tribute to King Louis XIV, the 'Sun King', who was currently on the throne. These two royal attributes can be envisaged if one regards the lower half of the constellation as being the Sceptre, and the upper formation the Hand of Justice, with its tip as Alpha Lacertae. But both royal commemorations were abandoned in the late seventeenth century when the German astronomer Hevelius introduced The Lizard: out with the regalia, in with the reptile!

Scheat (star Beta Pegasi)
Some astronomers have interpreted this name, which is certainly of Arabic origin, as a 'lucky' one, deriving it from Saʿd, 'good fortune', in the same way that such stars as *Sadachbia* and *Sadalmelik* contain this root word. But the general current view is that the name actually represents Arabic Al Sāʿid, 'the upper part' (of the arm), because this is the location that it marks on the outline of the horse that is the constellation *Pegasus*. (Compare the names of other stars here, which also mark parts of the figure, such as *Markab*, *Algenib* and *Enif*.) This seems to be confirmed by an alternative Arabic name for the star, which was Mankib al Faras, 'shoulder of the horse', formerly corrupted in English as Menkib.

Schedar (star Alpha Cassiopeiae)
This is often (as a possible variable star) the brightest star in the constellation of *Cassiopeia*, depicted sitting in her chair, and its

name derives from Arabic Al Sadr, 'the breast', for this is the point it marks on her figure. Compare *Sadr*.

Scorpion, The see **Scorpius**

Scorpius (constellation)
The name, also known as Scorpio, translates easily as The Scorpion, and is an ancient designation for this constellation, full of interesting stars and names. In classical mythology, the scorpion killed *Orion* with its sting, and in the night sky he can be seen fleeing before it still, because Orion sets below the horizon when Scorpius rises above it. For once, the outline does clearly resemble a fearsome scorpion, with its sting represented by the curve of stars that terminates in *Shaula* ('the sting'), Lambda Scorpii. The Greek name was thus Scorpios, and the Arabic, having the same meaning, Al ʿAḳrab. The heart of the creature is marked by its brightest star, *Antares*. See also *Libra*.

Sculptor (constellation)
The name means what it says, because the English word is taken directly from the Latin. Thus The Sculptor was the name of the constellation introduced in the 1750s here in the southern hemisphere by the French astronomer Nicolas Louis de Lacaille, who called it in full L'Atelier du Sculpteur, Latinised as Apparatus Sculptoris, 'the sculptor's workshop'. It is hard to see a resemblance of the stars it contains to a sculptor's workshop, or indeed to anything in particular, because the constellation is a faint and obscure one. But perhaps Lacaille intended the name to complement that of *Caelum*, The Chisel, as the latter will be needed in the former. However, the two constellations are nowhere near each other, so any connection of this type is purely speculative.

Sculptor's Chisel, The see **Caelum**

Scutum (constellation)
This constellation was introduced between *Aquila* and *Serpens* in the late seventeenth century by the German astronomer Hevelius, under the full name of Scutum Sobieskii, 'Sobieski's Shield'. This was a tribute he wished to pay to his patron, the Polish King John III Sobieski. The outline of the stars was taken to be the king's coat of arms, basically a shield with a cross, its centre marked by the faint constellation's brightest star, Alpha Scuti. The formation contains no named stars, however, and today is best known for its very unroyal-sounding *Wild Duck Cluster*.

Sea Monster, The see **Cetus**

Seething Bay see **Aestuum, Sinus**

Seginus see **Haris**

Serenitatis, Mare ('sea' on Moon)
The pleasant name means Sea of Serenity, and first appears on the Italian astronomer Riccioli's map of 1651, where it lies to the north of the similarly named *Mare Tranquillitatis*, in the north-east sector of the near side of the Moon. But although 'Sea of Serenity' is the usual English rendering of the Latin name, it should really be understood as 'Sea of Fair Weather'. As such, the name contrasts in the way favoured by Riccioli (or by his pupil Grimaldi) with some other 'stormy' or 'inclement' name. Perhaps this was the former adjacent 'Terra Nivis', or 'Land of Snow', a name now abandoned.

Serenity, Sea of see **Serenitatis, Mare**

Serpens (constellation)
The name, translating obviously as The Serpent, is an ancient one, and represents a serpent or snake encircling the body of *Ophiuchus*, The Serpent Holder. Unusually, the constellation itself is in two separate halves, one each side of Ophiuchus. The larger half, above him, is known as Serpens Caput, 'the serpent's head', while the other, below him, is called Serpens Cauda, 'the serpent's tail'. Ophiuchus is thus represented as grasping the Serpens Caput at his stars Delta and Epsilon Ophiuchi (otherwise *Yed Prior* and *Yed Posterior*), while his other hand grasps the Serpens Cauda at the stars Nu and Tau Ophiuchi. The serpent's actual head is depicted at the top of its long neck (marked by *Unukalhai*, Alpha Serpentis), where it is usually seen as formed by the three stars Beta, Gamma and Kappa Serpentis. The Greek name for the constellation was simply Ophis, with the same meaning, and the Arabic name also meant 'the snake', as Al Hayyah. This was a translation of the Greek name, however, and earlier the Arabs had seen quite a different depiction here, since they called the stars Al Raudah, 'the pasture'. This was a sort of field containing sheep, with the stars Beta and Gamma Serpentis, together with Gamma and Beta Herculis, called the Nasak Shāmiyy, 'the northern boundary', while Delta, Alpha and Epsilon Serpentis, together with Delta, Epsilon, Zeta and Eta Ophiuchi, were known as the Nasak Yamāniyy, 'the southern boundary'. The sheep were represented by the stars that

now form the Club of *Hercules* in his constellation, while the shepherd was the star Alpha Ophiuchi (*Ras Alhague*, now the head of the snake charmer), and his dog was Alpha Herculis (*Ras Algethi*, now the head of Hercules)!

Serpent, The see **Serpens**

Serpent Holder, The see **Ophiuchus**

Seven Sisters, The see **Pleiades**

Sextans (constellation)
The name means The Sextant, and it was introduced by the German astronomer Hevelius in the late seventeenth century. Hevelius used a sextant regularly for fixing star positions, and in fact continued to use it after most of his contemporaries had switched to making observations and reckonings by telescope. The constellation is a small and somewhat sparse one, but the angular outline of a sextant can perhaps be seen in its three main stars, Beta, Alpha and Gamma Sextantis, which, with the pivot at Alpha Sextantis, just about form the angle of 60°, or the arc of a sixth of a circle, that the sextant could measure at its maximum.

Sextant, The see **Sextans**

Seyfert galaxy
The name given to a spiral galaxy that has an unusually bright nucleus, such as M 77 in *Cetus*. Such galaxies were first identified in 1943 by the American astronomer Carl Seyfert, and are named after him.

Shaula (star Lambda Scorpii)
The star is located at the very end of the curving tail of The Scorpion who appears in the constellation of *Scorpius*. As such, it is therefore rightly named, because it represents Arabic Al Shaulah, 'the sting'. Understandably, the star was an unlucky one with astrologers.

Sheliak (star Beta Lyrae)
The name represents one of the alternative Arabic names for the constellation of *Lyra*, which was Al Shilyāk, 'the tortoise'. To some observers, the formation of stars here may even more readily suggest a tortoise than a lyre, if *Vega* is taken as its prominent head, and Gamma and Theta Lyrae as its back legs.

Sheratan (star Beta Arietis)
The name stands for Arabic Al Sharaṭain, 'the signs' (grammatically

the dual of Al Sharaṭ, 'the sign'), with this referring to both Beta and Gamma Arietis, whose appearance in the sky was regarded as marking the new year. The name is now used only of Beta Arietis, while Gamma Arietis is known as *Mesarthim*.

Shield, The see **Scutum**

Ship Argo, The see **Argo**

Sinope (satellite of *Jupiter*)
Sinope is the outermost of Jupiter's known satellites, and was discovered in 1914 by the American astronomer Seth Barnes Nicholson, who was much later to discover *Lysithea, Ananke* and *Carme*. Like the other names of Jupiter's satellites, Sinope, as a character in classical mythology, had a connection with Zeus (Jupiter). According to one story, Zeus took Sinope to the ancient city of this name on the Black Sea and promised to grant her dearest wish. She replied that she wished to remain a virgin, and thus managed to remain one for life. The city of Sinope (or Sinop) is still there on the Turkish Black Sea coast. (See also *Pasiphaë* for the significance of the final '-e'.)

Sirius (star Alpha Canis Majoris)
Sirius, the Dog Star, is not only the leading star in *Canis Major*, The Greater Dog, but also the brightest star in the whole sky. Its name comes from Greek Seirios astēr, 'the scorching star', with this referring not only to its brilliance but also to the fact that it was near the Sun during the hottest season of the year, midsummer, and that it thus added to the heat already given by the Sun. The Romans borrowed the name and the concept, calling these midsummer days *caniculares dies*, 'dog days', and this in turn gave the English phrase 'dog days', either for this midsummer period, when dogs were also (wrongly) supposed to be liable to hydrophobia, or for any period of inactivity. The Latin phrase also gave the standard Russian word for 'holidays' of the school type, which is *kanikuly*. The school summer holidays, after all, take place when the weather is at its hottest (in theory), and at a time when it is thus difficult to work. The Arabic name for the star was identical, with Al Shiʿrā meaning 'the shining one'. The full name of the star was also the imposing Al Shiʿrā al ʿAbūr al Yamāniyyah, 'the shining one of the passage of Yemen', because Sirius had a southern position in the sky, in the direction of Yemen (whose own name means 'right', that is, to the right of a Muslim as he faces Mecca when praying). In this respect Sirius was different from *Gomeisa*,

145

in *Canis Minor*, which was visible in the north, towards Syria (whose own name means 'left' for the contrary reason). Compare *Procyon*.

Sirrah see **Alpheratz**

Skat (star Delta Aquarii)
The name is almost certainly of Arabic origin, although its precise sense remains uncertain. It may be a corruption of Al Sa'd, 'the good fortune', and so be a name on the same lines as those of Alpha, Beta and Gamma Aquarii, respectively *Sadalmelik*, *Sadalsuud* and *Sadachbia*, all 'lucky' names. Alternatively, from its position on one of the legs of the figure who is The Water Carrier of the constellation, it may derive from Al Şāk, 'the shinbone'. Clearly a corruption of the original has taken place either way.

Smythii, Mare ('sea' on Moon)
The *mare* is on the eastern edge of the near side of the Moon, where it extends round to the far side. It was named in honour of the English admiral and astronomer, William Henry Smyth (1788–1865), who had a private laboratory at Bedford and who wrote both *The Sailor's Word-Book*, as a sort of dictionary of seafarers' jargon, and *The Cycle of Celestial Objects for the use of Naval, Military, and Private Astronomers*, thus contributing his experience in both his professional roles. (One of his sons, Charles Piazzi Smyth, named after the Italian astronomer, himself became Astronomer Royal for Scotland.)

Smyth's Sea see **Smythii, Mare**

Sobieski's Shield see **Scutum**

Sombrero Hat Galaxy (in *Virgo*)
The nickname for the galaxy officially designated NGC 4594, or M 104 (in Messier's catalogue), is descriptive of its distinctive appearance when seen on long-exposure photographs. It can be observed edge-on like this, and looks something like the planet *Saturn* encircled by its rings horizontally, or again it also conjures up a flying saucer (of the science fiction type). But instead it has earned its present name, comparing it to a sombrero hat.

Somnii, Palus (feature on Moon)
The name translates as Marsh of Sleep, and the feature is really a small 'sea' leading off the *Mare Tranquillitatis*. On Riccioli's original map of 1651, it lay close to the former 'Terra Vitae', or 'Land of Life', and so was one of the pleasant or 'healthful' names in this area of the Moon. For a similar name, see *Lacus Somniorum*.

Somniorum, Lacus (feature on Moon)
This name translates as Lake of Dreams, and is in effect a small 'sea' or *mare* leading off the larger *Mare Serenitatis*. It is thus typical of this region of the near side of the Moon, where there are a number of 'restful' names. However, lying to the northern side of the Lacus Somniorum is the *Lacus Mortis*, as if the further north one goes, the more somnolent and 'frozen' one becomes.

Southern Cross, The see **Crux**

Southern Crown, The see **Corona Australis**

Southern Fish, The see **Piscis Australis**

Southern Fly, The see **Musca**

Southern Lights, The see **Aurora Australis**

Southern Sea, The see **Australe, Mare**

Southern Triangle, The see **Triangulum Australe**

Spica (star Alpha Virginis)
Many legends connect *Virgo*, The Virgin, with an abundant harvest, and when she is regarded in this way, she can be seen in the constellation holding an ear of wheat in her left hand (and a palm leaf in the other). This ear of wheat is the star Spica, whose Latin name means just that. (Compare the English word 'spike' in the sense of 'ear of corn'.) The former Greek name for the star was Stachys, with exactly the same meaning. The Arabic name for it was Sunbala, deriving from Al Sunbalah, 'the sheaf of wheat'. This was the original name they used for the constellation as a whole, because they would not depict the human figure.

Stern, The see **Puppis**

Storms, Ocean of see **Procellarum, Oceanus**

Sualocin (star Alpha Delphini)
Not an Arabic name, despite its exotic appearance, but a simple reversal of the name Nicolaus! This was the Latinised first name of Nicolaus Venator, the Italian astronomer whose native name was Niccolo Cacciatore and who was assistant to Giuseppe Piazzi. The two star names of Sualocin and *Rotanev* first appeared in Piazzi's catalogue of 1814, and were given to the two brightest stars in *Delphinus* as a tribute to Cacciatore (alias Venator). The enjoyable thing is that both new names fooled more than one astronomer

into interpreting them as Arabic names, and the nineteenth-century writer Frances Rolleston, in her strange book *Mazzaroth*, derived Sualocin from 'Arabic Scalooin, *swift (as the flow of water)*' and Rotanev from 'Syriac and Chaldee Rotaneb, or Rotaneu, *swiftly running (as water in the trough)*'! But one must not be too hard on Miss Rolleston, because such formations were not normally the order of the day, even in 1814, when it came to naming new stars. (The title of her book is a biblical name, found in Job 38: 32: 'Canst thou bring forth Mazzaroth in his season?'. This means 'signs of the zodiac', and as a name has been applied to various stars and star groups over the centuries. The biblical line is preceded by the better known: 'Canst thou bind the sweet influences of Pleiades, or loose the hands of Orion?')

Sulafat (star Gamma Lyrae)
The name is of Arabic origin, representing Sulahfat, 'tortoise', as this was the animal they saw depicted here, with this name used for the constellation as a whole. They were not alone in this, because a Roman name for the grouping was Testudo, with the same sense. The outline of a tortoise can be clearly made out in the sky here, with *Vega* as its head. In fact there is an old legend that links both tortoise and lyre, because the original instrument was imagined as being made out of a tortoise shell, especially one of a creature that had died and that had been washed up upside down on the beach, with the dried tendons of the animal making the strings of the lyre. See also *Sheliak*.

Summer Triangle (star group in *Lyra* and neighbouring constellations)
An unofficial name for the three bright stars forming a triangle in the summer night sky: *Vega* marks one angle, and the other two are marked by *Deneb* (in *Cygnus*) and *Altair* (in *Aquila*). The name was introduced by the British astronomer Patrick Moore in one of his television programmes in the late 1950s. In the southern hemisphere, of course, the three stars would appear as a 'Winter Triangle'.

Sun, The (star at centre of Solar System)
The name of our own lightening and life-giving star is similar in many languages, often beginning with 's-' and frequently containing an 'n', especially in Germanic and Slavonic languages. We thus have French *soleil*, Italian *sole*, Spanish *sol* and Portuguese *sol* on the one hand, and German *Sonne*, Dutch *zon* and Russian *solntse*

(but Swedish *sol*) on the other. All these are directly related to Latin *sol*, Greek *helios* and Welsh *haul*, which in turn are related to one another. The initial 's-' probably means 'shine', in which English word it is also seen (as well as in 'sheen', 'sheer' and 'shimmer'). The Sun is thus a 'shiner'. (The relationship between the words is even more marked in some languages, such as Russian *svet*, 'light', *svetit*, 'to shine' and *svetilo*, 'celestial body', 'luminary'.)

Swan, The see **Cygnus**

Swordfish, The see **Dorado**

Sword Handle (star group in *Perseus*)
The nickname is sometimes used for the two star clusters that appear near the point where Perseus holds his outstretched sword. They are also simply known as the Double Cluster, and are often designated as h and Chi Persei. The Arabs called them Miʿsam al Thurayya, 'wrist of the *Pleiades*', because they are near this famous group (in neighbouring *Taurus*).

Table, The see **Mensa**

Tarantula Nebula (in *Dorado*)
The nebula, officially catalogued as NGC 2070, is nicknamed for its spidery shape in the Large *Magellanic Cloud* when observed through a telescope. It is the largest nebula known at present.

Tarazed (star Gamma Aquilae)
The name is part of the Persian title of the whole constellation of *Aquila*, which was Shahin tara zed, 'the star-striking falcon'. The first half of this gave the present name of *Alshain*, Beta Aquilae. Both these stars can now be seen 'flying' either side of The Eagle itself, *Altair*, Alpha Aquilae.

Taurids (meteor shower)
The name indicates that the annual shower, a minor one, emanates from the constellation of *Taurus*. The shower can be observed in October and November, with its peak period in the first half of the latter month.

Taurus (constellation)
Taurus is an ancient name, The Bull, for a constellation that has long been popularly associated with strength and fertility. As depicted in the sky, it is the bull's head, with horns, that is seen, with its face formed by the cluster of stars called the *Hyades* and

its glittering eye marked by its brightest star, the red giant *Aldebaran*. Its horns extend outwards to Beta and Zeta Tauri, the former known appropriately as *El Nath*, 'the butting one'. The constellation is generally believed to have been the first that primitive man recognised and distinguished in the heavens. The Greek name for it was Tauros, to which the present Latin name is identical. The Arabic name, also meaning 'the bull', was similarly Al Thaur. In classical mythology, the particular bull of the constellation is usually identified with the one with brazen feet that was tamed by Jason. However, there must surely be also a link between the famous (or infamous) Minotaur, the monster that was half man, half bull. This creature's own name was Asterion, 'starry', and when in his labyrinth, he was annually fed seven young men and seven girls. Today the constellation of Taurus contains two familiar seven-star groups, the *Pleiades* (Seven Sisters) and the Hyades already mentioned above. Although both groups are now thought of as representing seven girls, it is perfectly possible that the Minotaur story simply converted one group to young men for the sake of artistic or narrative 'balance'. Either way, the constellation and its contents is one of the most potent and sexually charged that the night sky holds, with its contrasting symbolism and imagery of strong versus weak, male against female, one (The Bull) enclosing or 'devouring' many (the Pleiades and the Hyades).

Telescopium (constellation)
A modern name for a constellation that is still appropriate today, among the outmoded *Fornax, Antlia, Octans* and *Horologium*. It was introduced, as were all these, by the French astronomer Nicolas Louis de Lacaille in the 1750s. Alas, the constellation itself is unremarkable, even insignificant, with its resemblance to a telescope a purely contrived one, and owners of even the most basic telescope will find little or nothing of interest in it today to observe. It is located next to its 'opposite number', *Microscopium*.

Tethys (satellite of *Saturn*)
Tethys was discovered by the Italian astronomer Giovanni Domenico Cassini in 1684, the same year that he discovered *Dione*. Its name, like those of some of its fellow satellites, is found in classical mythology as one of the *Titans*, with Tethys being an ancient goddess of the sea. In some older astronomical works and maps, Tethys is misnamed as 'Thetis', doubtless through the similarity of the two names. In mythology, however, Thetis is quite a different character, as a daughter of Nereus and Doris. But she

was similarly a sea goddess, which probably added to the confusion of identity.

Thebe (satellite of *Jupiter*)
Thebe was discovered only in 1979, between the orbits of *Amalthea* and *Io*, when photographs taken by the space probe Voyager 2 as it flew by *Jupiter* were examined. In Greek mythology, Thebe was (among several other Thebes) the daughter of Zeus (*Jupiter*), so the satellite has a 'family' link, as do the names of most of the planet's other satellites. However, Thebe should not be confused with the satellite *Phoebe*, belonging to *Saturn*!

Themis (asteroid 24)
Themis was discovered in 1853 by the Italian astronomer Annibale de Gasparis, who earlier had also discovered many other asteroids, including *Hygeia*. In classical mythology, Themis was one of the *Titans*, the daughters of *Uranus* and Ge, and the second wife of Zeus (*Jupiter*). As such, her name would have been suitable for one of Jupiter's satellites; but it is not included among them. However, it once was: in 1904 the American astronomer William Henry Pickering claimed that he had discovered a new satellite of Jupiter, between the orbits of *Titan* and *Hyperion*. It was duly named Themis, and welcomed to the family. Subsequently, alas, it has been proved not to exist, so remains as a 'ghost' name for one of Jupiter's attendants. Fortunately it was already in use anyway. Fortunately, too, Pickering's other discoveries, such as *Phoebe* in 1898, were real enough. See also *Pluto* for pioneering work done by him that paved the way to the planet's discovery.

The NORC (asteroid 1625)
The name is included here as something of a curiosity, complete with its obligatory definite article and acronym. The asteroid was discovered in 1953 by the Belgian astronomer Sylvain Arend (after whom the Arend-Roland comet is named). The asteroid was called after the initials of the electronic instrument used for detecting it, the Naval Ordnance Research Calculator. The definite article is unique in asteroid names. Arend's own name is also preserved in that of minor planet 1502, Arenda.

Theophilus (crater on Moon)
The name is typical of the hundreds of names of scientists and philosophers with which many features of the Moon are commemorated. This crater, a large one on the edge of the *Mare Nectaris*, was named after the second-century saint who was bishop

of Antioch and an ecclesiastical writer. It was given to the crater by the seventeenth-century Belgian astronomer and globe-maker, Jacob van Langren, who similarly 'baptised' many other craters on the Moon. According to an article published in 1960 by the Polish researcher Stanisław Brzostkiewicz, only three of Van Langren's original names now remain: Theophilus, Cyrillus and Catharina. The latter two, also named after saints (Cyril and Catherine), are craters close to Theophilus.

Thuban (star Alpha Draconis)
The Arabic name of this star is usually said to derive from Al Thu'bān, 'the dragon' and thus to refer also to the constellation of *Draco* as a whole. But this name appears to be a misinterpretation made in the sixteenth century by the French historian Joseph Scaliger, when he altered an already miscopied Arabic word *tabbīn* to *Thuban*. The earlier miscopying, made by European translators, was of the original Arabic *tinnīn*, another word for 'dragon'. So the true origin of the name is in Arabic Al Tinnīn, 'the dragon'. Fortunately, although the word has been corrupted more than once over the centuries, the original sense has been retained. (The same error occurred in the name of *Rastaban*. See *Alwaid*.) Somewhat surprisingly, in its role as constellation symboliser, the star is not the most prominent in *Draco* in either its brilliance or its position. It is not in the dragon's head, but halfway down its body. Its rightful position is thus occupied by *Eltanin*, Gamma Draconis. However, it once was the Pole Star (about 5000 years ago), until it lost this prominent place to *Polaris* because of the effects of precession (see Glossary), so has had its day of importance, even if it was no brighter then than now.

Titan (satellite of *Saturn*)
Titan is Saturn's largest satellite, and its name is appropriate for its prominence, because many of its fellow satellites bear the names of the Titans (or Titanesses) in Greek mythology, where they were the giant first-born children of *Uranus* and Ge, and brothers of Cronos (Saturn), their ruler. Not surprisingly, on account of its size, Titan was the first of Saturn's satellites to be discovered, and this fell to the Dutch astronomer Christiaan Huygens in 1655. He can be criticised for giving the satellite a 'generic' name, not a specific one (of an individual character in mythology), but at the same time he established a naming pattern for Saturn's satellites that could be followed by the selection of individual Titans for the names of subsequent discoveries.

Titania (satellite of *Uranus*)
Titania and *Oberon* were the first two satellites of Uranus to be discovered, and when William Herschel did so in 1787 he named both after the two main characters, king and queen of the fairies, in Shakespeare's *A Midsummer Night's Dream*. Herschel clearly felt that most of the standard mythological names had been used up – or preferred the stories of Shakespeare to the legends of the classical authors! At any rate, he struck a new path in naming, which was followed for many of the planet's other satellites, discovered subsequently, among them *Miranda*. It does not seem to have worried Herschel that there was no connection at all, literary or otherwise, between his chosen names and that of their host planet. Indeed, just the reverse, because Uranus (Cronos) fathered monsters and giants (the Cyclopes and the *Titans*), whereas Titania and Oberon presided over much smaller, gentler, albeit mischievous beings.

Toucan, The see **Tucana**

Tranquillitatis, Mare ('sea' on Moon)
The name is usually translated as Sea of Tranquillity, and is that of one of the major *maria* or 'seas' on the Moon's surface. It appears on Riccioli's lunar map of 1651, where it is one of the 'restful' names found in this eastern half of the Moon, among others such as the *Mare Serenitatis*, *Palus Somnii* and *Lacus Somniorum*. It was in the Mare Tranquillitatis that the first lunar landing was made, when the American crew of Apollo 11 touched down in 1969 (at the 'Tranquillity Base'). The name seemed symbolic to many at the time for the need to explore and exploit the Moon in a calm and peaceful manner. But what if the spacecraft had landed in the *Mare Crisium*?

Tranquillity, Sea of see **Tranquillitatis, Mare**

Trapezium (star Theta Orionis)
The name is really a nickname for this multiple star, for even a fairly basic telescope will show it to consist of four stars in the shape of a trapezium, that is, at the four corners of the mathematical figure of this name having two parallel sides. The star is in the southern sector of the *Orion* Nebula. It seems strange that no ancient names have been recorded for the four stars, either individually or collectively, especially as the Trapezium marks Orion's sword as it hangs from his belt. (But see *Saiph*!)

Triangle, The see **Triangulum**

Triangulum (constellation)
The name is really an ancient one, not a modern scientific or
mathematical introduction, although the Greeks called the constellation either Deltoton (after the triangular capital Greek letter
delta) or Trigonos ('three-angled' or 'triangular'). This latter name
was Latinised by the Romans as Trigonum, and the present Latin
equivalent appeared in more recent times. It was also called Sicilia
by them, because Sicily is a three-sided island (roughly). The Arabic
name, Al Muthallath, also means 'the triangle'. Needless to say,
the constellation's three main stars do actually form a delta shape,
although individually Alpha, Beta and Gamma Trianguli have no
names of their own, apart from the first of these, which is at the
'point' and so is sometimes called Caput Trianguli, 'head of the
triangle' translated by the Arabs as Rās al Muthallath (Rasalmo-
thallah). Compare *Triangulum Australe*.

Triangulum Australe (constellation)
Here is another triangle to match the one seen in *Triangulum*. As
its name indicates, it is in the southern hemisphere, hence the
second descriptive word of the title, which translates as Southern
Triangle. The name was introduced in 1603 by the German astronomer Johann Bayer. It is one of his more prosaic creations, because
he usually went in for the names of exotic birds and animals, such
as *Apus*, *Chamaeleon*, *Grus*, *Pavo* and *Tucana*.

Trifid Nebula (in *Sagittarius*)
The name is nothing to do with the hostile mobile plants of John
Wyndham's science fiction classic, *The Day of the Triffids*. Etymo-
logically, however, the two words are probably related, and the
Trifid Nebula is so called because it is trisected by three dark lanes
of dust. (The '-fid' means 'split', from the base of Latin *findere*, 'to
cleave'. The fictional Triffids appear to have a name based on
'Trifid', because they are described as being supported on 'three
bluntly-tapered projections'.) The Triffid Nebula is officially known
as M 20 or NGC 6514.

Triton (satellite of *Neptune*)
Triton is the larger of *Neptune*'s two satellites, and was discovered
in 1846 by the English astronomer William Lassell less than three
weeks after the discovery of Neptune itself. Its name is perfectly
apt, as in classical mythology Triton was the minor sea god who
was the son of Poseidon (Neptune). The planet's other satellite,

Nereid, was discovered over a hundred years later. Lassell was also the co-disoverer and sole namer of *Hyperion*, one of the satellites of *Saturn*.

Trojans (group of asteroids)
The Trojans are those asteroids that move in virtually the same orbit as *Jupiter*, either ahead of this planet (the 'Leading Trojans') or behind it ('Trailing Trojans'). They are so called because they are named after the heroes of the Trojan War in classical mythology, with the brightest asteroids of the group called Achilles, Patroclus, Hector, Nestor, Priamus, Agamemnon, Odysseus, Aeneas, Anchises, Troilus, Ajax, Diomedes, Antilochus, Menelaus and Telamon. (It should be noted that some of these names are not of Trojans but of Greeks, such as Agamemnon and Menelaus.) The masculine names indicate the unusual orbits of the asteroids (see p. 235).

Tucana (constellation)
The name is Latin for The Toucan, and as might be expected was one of the exotic bird and animal names introduced by the German astronomer Johann Bayer in 1603. The outline of stars in the constellation does actually resemble the American bird, with its distinctive large beak (formed by Beta, Zeta and Epsilon Tucanae). Its eye can be seen as Gamma Tucanae.

Twins, The see **Gemini**

Tycho (crater on Moon)
Tycho is one of the most prominent ray craters on the Moon's surface, in the south-west sector of the near side. It appears on Riccioli's map of 1651, based on that of his pupil Grimaldi, and is an example of the latter's wish to commemorate leading men of learning, especially astronomers. Tycho was the sixteenth-century Danish astronomer Tycho Brahe, one of the greatest of the pre-telescopic celestial observers and star cataloguers. He was, though, still convinced that the Earth was the centre of the Solar System.

Umbriel (satellite of *Uranus*)
Of the five chief satellites of Uranus, Umbriel and Ariel are the only ones not to derive from Shakespearean characters. Umbriel was discovered by the English astronomer William Lassell in 1851, the same year that he discovered *Ariel*. (In fact it had probably been glimpsed earlier by Herschel, but it was Lassell who now recovered and 'captured' it.) Lassell named both satellites, with

Umbriel 'a dusky melancholy sprite' in Pope's poem *The Rape of the Lock*. Pope seems to have based the name on Latin *umbra*, 'shade', so that astronomically the name happens to be apt, because Umbriel can be regarded as 'shadowing' Uranus, as one of its satellites.

Undarum, Mare ('sea' on Moon)

The *mare*, whose name means Sea of Waves, is near the *Mare Crisium*, and not far from the *Mare Spumans*, so has a suitable 'stormy' name for this part of the near side of the Moon. The name appears on Riccioli's seventeenth-century lunar map.

Undina (asteroid 92)

Undina was discovered in 1867 by the Danish astronomer Christian Peters, who also discovered nearly fifty other asteroids. The name is characteristic of a run of asteroid names at about this time, when classical mythology was temporarily abandoned in favour of names from other legends. In this case, Undina was the name created by the highly inventive and imaginative Swiss alchemist Paracelsus (who also invented his own name), for a special sort of sea nymph. (He based the word on Latin *unda*, 'wave'.) Since his time, 'undine' has become a generic term for a nymph of this type, and as an individual supernatural female spirit has become the subject of various literary and musical works, such as La Motte Fouqué's tale 'Undine' and the opera based on it. Possibly one of these works was the direct inspiration for the asteroid name, rather than Paracelsus's original nymph. Other names from non-classical mythology include asteroids 76 Freia and 77 Frigga, both from Norse legend. The latter was also discovered (in 1862) by Peters.

Unicorn, The see Monoceros

Unukalhai (star Alpha Serpentis)

The name comes from Arabic 'Unk al Hayyah, 'the neck of the snake', because this is the part of the snake who is *Serpens* that the star marks (in the half of the constellation called Serpens Caput, where it is the brightest star in the whole constellation).

Uranus (planet)

The seventh planet of the Solar System was discovered in 1781 by the English (but German-born) astronomer William Herschel (who at first thought it was a comet). He may have been an excellent astronomer, but Herschel was not a very effective namer, and at first he wished to call the new planet 'Georgium sidus', Latin for

'George's star', as a tribute to the reigning monarch, George III. But such a name would have broken with all accepted conventions, and it was unheard of to have a major celestial body named after a king, to say nothing of the fact that it was not a star (which is the prime meaning of Latin *sidus*, even though it can also mean 'heavenly body'). However, when the Finnish astronomer Anders Lexell was addressing a learned audience in St Petersburg, Russia, two years later, in order to develop his theory that Herschel's discovery was indeed a planet and not a comet, he proposed a name that was hardly any better, offering 'George III's Neptune', or 'Britain's Neptune'. He intended this to be a compliment to the victories of the British navy in the two previous years, during the American Revolution. After all, *Neptune* was the god who symbolised the sea! His suggestions were rejected, because they also broke with tradition, and were likewise too lengthy. Nor was another name proposal accepted: that made by the French astronomer Joseph Lalande, who offered the name 'Herschel' for the planet. This was more appropriate than the name of the king, but once again there was no precedent for naming a major celestial object after a living astronomer. Meanwhile, Swedish astronomers proposed 'Neptune' on its own, and Lexell himself agreed with the suggestion. It was the German astronomer Johann Bode, however, who pointed out that Uranus, as a classical name, would be more appropriate than Neptune, even though the latter correctly followed the accepted mythological pattern. The sixth planet in order from the Sun was called by its ancient Roman name of *Saturn* (the Greek god Cronos), so what was needed now for the seventh planet was a name that directly linked with this in a genealogical way, just as Saturn's linked with *Jupiter*, the fifth planet. (Cronos was the father of Zeus, that is, Jupiter.) It was thus right to name the seventh planet Uranus, because in mythology he (the personification of the sky) was the father of Cronos. There would therefore be a direct 'family line' running from son up to father up to grandfather. The only way in which the name did somewhat differ from those of the other planets, however, was that it was a Greek name. All the other planets have the names of Roman gods and goddesses. But this was unavoidable, because the Romans did not have a name for a god that corresponded to Uranus. And in the end Neptune did find a home on a planet, as we know. See his name for the reasoning behind it. (Uranus himself had no father, only a mother, the earth goddess Ge.)

Ursa Major (constellation)
This ancient constellation, The Great Bear, is familiar even to
non-astronomers because of its prominent star formation known
variously as the Plough, or the Big Dipper, or Charles's Wain, or
the Northern Wagoner, or simply the Seven Stars, among other
names. ('I prefer "The Bent Saucepan" ', says a frivolous friend.)
The main stars of the constellation can indeed be envisaged as a
large bear, stalking across the sky, with its nose at Omega Ursae
Majoris, and the tip of its tail at *Benetnasch*, Eta Ursae Majoris.
But this last name is a reminder that different peoples in the past
have seen a different figure represented here, in particular a coffin.
(A bier, not a bear!) But similarly the reference to a bear is
preserved in the name of *Dubhe*, Alpha Ursae Majoris, its brightest
star (midway along the bear's back). The Greek name for the
constellation, or more exactly for its seven main stars, was Arctos,
'the bear', while the Arabs knew it as Al Dubb al Akbar, 'the
greater bear'. But this second name crystallised only when *Ursa
Minor* had been established as 'the lesser bear'. What bear was it
that the original namers saw here? With which particular animal,
real or legendary, did they associate it? We know that another
Greek name for the formation was Callisto, and the story that
accounts for this is the one in which the nymph of this name,
having vowed to remain a virgin, spent her life hunting with the
companions of Artemis (the Roman Diana, goddess of the hunt).
Zeus (Jupiter) saw her and seduced her disguised as Artemis, for
Callisto shunned all men. As a result of the union she became
pregnant with a son named Arcas, and her condition was revealed
when she had to undress one day to bathe in a spring with (the
real) Artemis and her other companions. As a punishment for
breaking her vow of virginity, Artemis changed Callisto into a she-
bear, represented by the present constellation. (A variant of the
story tells how Zeus similarly changed her into the constellation.)
The name Arcas itself appears to have a close link with Greek
arktos, 'bear'. But the constellation name predatès even the
Greeks, and one can only guess at the ultimate inspiration for the
designation. All one can say with certainty is that the configuration
of the stars, thousands of years ago, would have even more vividly
suggested a bear than they do now, and that having noted and
preserved this depiction, our remote ancestors invented various
tales and legends to commemorate it. This appears to have
happened independently in different parts of the world, so that
stories about the great bear here have been found in Asia and North

America, as well as in Mediterranean lands. It is still uncertain when the name was extended to apply to the constellation at its fullest extent, beyond the original seven, as we know it today. It was certainly before the formation of Ursa Minor, which is itself an ancient name, although datable with greater precision. The original 'magnificent seven' are actually (working back from the middle of the bear's body to the tip of its tail) the stars Alpha, Beta, Gamma, Delta, Epsilon, Zeta and Eta Ursae Majoris, otherwise respectively Dubhe, *Merak, Phecda, Megrez, Alioth, Mizar* and Benetnasch (or *Alkaid*). These last three form the 'handle' of the Plough or the Big Dipper (the ladle). But the seven have also been seen as a wagon or chariot, and this has given the two English names of Charles's Wain and Arthur's Wagon. The first of these (more common than the second) pays tribute to Charlemagne ('Charles the Great'), while the second is associated with King Arthur. Not for nothing does the name Arthur link up with Greek *arktos*, 'bear', just as Arcas, already mentioned, does, as well as the word for the animal in Celtic languages. (The Welsh for 'bear' is simply *arth*, for example.) So the constellation as a whole is full of imagery, allusion and 'mystery and history', and far more names exist for it, and for its prime seven stars, than can possibly be mentioned here. (The Greater Bear, incidentally, is the one that *Boötes*, The Herdsman, pursues across the sky. And it was the Greek name of the constellation, Arctos, that gave the name of the Arctic as the north polar region, although it is Ursa Minor that is centred on the north celestial pole.)

Ursa Minor (constellation)
The constellation of Ursa Minor, The Lesser Bear, is said to have been introduced in about the year 600 BC by the Greek astronomer Thales, by contrast (and comparison) with the much larger *Ursa Major*. The Greeks also knew the constellation as Cynosouros, literally 'dog tail', although the name Cynosura came subsequently to refer solely to its brightest star, better known as *Polaris*. The question is, though: the tail of which dog? It can hardly be the Dog Star that is *Sirius*, in *Canis Major*, because this constellation is in the southern hemisphere. And although there are other dogs in the sky, as in *Canes Venatici*, there would appear to be no connection with them, either. One can only assume that, unless there has been some mistranslation or corruption of the name, the actual 'tail' comprises the stars that lead to the Pole Star, with the rest of the animal at some time thought of as a dog, instead of a bear. Possibly

mariners thought of the formation in this way, when using Polaris to steer their ships. The use of Polaris as a 'guiding star' in fact lies behind the modern word 'cynosure', meaning 'person or thing that attracts notice' ('Where perhaps some beauty lies, The cynosure of neighbouring eyes', Milton). The Pole Star was the one that attracted and 'drew' the navigators, and the general sense developed from this idea. An alternative English name for the seven brightest stars of Ursa Minor is the Little Dipper, because in a way their formation resembles that of the Big Dipper in Ursa Major, with Polaris at the end of the 'handle'. The Arabs knew the constellation as Al Dubb al Aşghar, 'the lesser bear', although just as they had seen a coffin in Ursa Major, they also fancied one depicted here, and so called the three stars in the 'tail' Banāt al Na'ash al Şughrā, 'the daughters of the lesser bier'. (Compare the name of *Benetnasch* or *Alkaid* in the larger constellation.)

Ursids (meteor shower)
The annual shower of meteors radiates, as its name implies, from one or other of the constellations *Ursa Major* or *Ursa Minor*. The trouble is, we cannot tell which, without further information. Astronomers will know, however, that they emanate from Ursa Major, appearing in late December.

Van Allen Belts (regions of charged particles above Earth)
The Van Allen Belts (or Van Allen Zones) contain charged particles resulting from cosmic rays, and are trapped at two different altitudes by the Earth's magnetic field. (The inner belt, containing mainly protons, is at about 2400 to 5000 kilometres; the outer, containing mainly electrons, is from 13,000 to 19,000 kilometres.) They are named after the American physicist who discovered them in 1958, James A. Van Allen.

Vaporum, Mare ('sea' on Moon)
The Mare Vaporum, or Sea of Vapours, is a small dark *mare* near the centre of the Moon's surface, as seen from Earth. The name appears on Riccioli's mid-seventeenth-century lunar map, where it is one of the 'watery' designations in this region, roughly between the *Mare Imbrium* and *Mare Nubium*.

Vapours, Sea of see **Vaporum, Mare**

Vega (star Alpha Lyrae)
The name looks Greek or Latin, but is actually Arabic in origin. It derives from the last part of the name for the constellation of

Lyra as a whole, which was Al Naṣr al Wāḳiʿ, 'the swooping eagle' (literally 'the eagle the falling one'), because the Arabs associated the star outline here with that of a bird, not a lyre. The name is thus in contrast to that of *Aquila*, which we do call The Eagle. The Arabic name for this was Al Naṣr al Ṭā'ir, 'the flying eagle' ('the eagle the flying one'), and it was the last part of this name that gave that of *Altair*, Aquila's most prominent star. Because the two constellations of Aquila and Lyra were almost adjacent in the sky, the Arabs eventually dropped the common 'eagle' element in both names, and left the defining words.

Veil Nebula (in *Cygnus*)
The name, which refers to the appearance of the nebula, is applied to the brightest part of the swirls of gas known as the *Cygnus Loop*. The nebula is probably the remains of a supernova (a so-called 'supernova remnant').

Vela (constellation)
The name means The Sails (as the plural of Latin *velum*), and was one of the four new constellations into which the former large *Argo* was divided by the French astronomer Nicolas Louis de Lacaille in the middle of the eighteenth century. The brightest stars here could be regarded as depicting the spread sails of a sailing ship. (There are no stars Alpha or Beta Velorum because the constellation is a subdivision of the larger Argo.) Two stars of the constellation, together with two in neighbouring *Carina*, form the *False Cross*.

Venus (planet)
Venus is the second planet from the Sun, and the only one to be named after a Greek goddess (as distinct from a Greek god). The original Greek names for Venus were Phosphoros (literally 'bearing light', that is, when seen in the morning) or Hesperos ('evening', when seen at this time of day: the name is related to English 'vespers'). This meant that Venus was regarded as two separate bodies, and is the origin of the poetic names 'Morning Star' and 'Evening Star' for the planet. It was Pythagoras, however, who first proposed the theory that the two apparent bodies were actually one and the same. And as we now know, Venus is not a star at all in the proper sense. The Greeks also knew Venus as Eosphoros. This means 'bearing the morning', and is a complementary name to the 'evening' one, while formed in the same poetic way as Phosphoros. Later, as for the other anciently named planets, Aristotle renamed the now unified Hesperos/Phosphoros as

Aphrodite, after the Greek goddess of beauty and love. The Romans in turn called the planet by the name of their own goddess who corresponded to Aphrodite. However, the Romans, too, had earlier given Venus a 'double' name, just as the Greeks had. Their equivalents of the 'morning' and 'evening' names were Lucifer ('bearing light') and Vesper ('evening'). Lucifer may seem a strange name in the circumstances, because it is the name of the Devil. But this latter use of the name came later, and refers to the biblical story of how the King of Babylon boasted that he would ascend to the heavens and be equal to God as a 'day star', but who was fated to be cast down to hell. (The story is told in Isaiah 14: 12 which reads: 'How art thou fallen from heaven, O Lucifer, son of the morning! how art thou cut down to the ground, which didst weaken the nations!')

Vesper see **Venus**

Vesta (asteroid 4)
Vesta was the fourth asteroid to be discovered. The astronomer concerned, who made his find in 1807, was the German Heinrich Olbers, who had earlier also discovered *Pallas*. In classical mythology, Vesta was the Roman goddess of the hearth, corresponding to the Greek Hestia (whose own name was subsequently given to asteroid 46 in 1857).

Victoria (asteroid 12)
Victoria was discovered in 1850 by the English astronomer John Russell Hind, who also discovered ten other asteroids, including *Iris* and *Flora*. Queen Victoria had by then been on the British throne for thirteen years, and Hind named the asteroid after her. This was a fine loyal tribute, but it was also a break with tradition, because up to this point no asteroid had been named after a monarch, living or dead. A number of Hind's colleagues thus objected to the name, and proposed instead 'Clio', after the Greek muse of history. But Victoria won the day, and Clio was in due course the name given to asteroid 84, discovered in 1865. (For another controversy concerning a royal astronomical naming, see *Uranus*.)

Vindemiatrix (star Epsilon Virginis)
The Latin name means 'gatherer of the vintage' (in the feminine form), and was a translation of the earlier Greek name, Protrygater (literally 'herald of the vintage'). The name both referred to the general association of *Virgo* with the harvest while indicating a

star that rose just before the grape harvest began. Compare the somewhat similar name of the star *Bellatrix*, Gamma Orionis, only there the association is with war, not with wine.

Virgin, The see **Virgo**

Virgo (constellation)
Virgo, The Virgin, is the second largest constellation in the sky, and the name is an ancient one, associated with many myths and legends, all centring on a beautiful and virtuous maiden. A particular association, however, was with the harvest, whether of wine or corn, and as depicted in the sky, Virgo holds an ear of corn in her left hand, as the star *Spica*. She is thus usually identified with the Greek goddess of the harvest, Demeter (whose Roman equivalent was Ceres), but there are also tales linking the constellation with another maiden, Persephone (the Roman Proserpine), similarly associated with the successful planting and harvesting of crops. In yet another capacity, The Virgin is thought of as Astraea, the Roman goddess of justice, with her scales in the neighbouring constellation of *Libra*. The basic Greek name for the constellation, corresponding to the present Latin one, was thus Core, 'the maiden', while the Arabs, at first depicting parts of the formation as a lion or as dogs barking at it, similarly adopted the Greek name and called the grouping Al ʿAdhrāʾ al Naṭhīfah, 'the innocent maiden'. But although they could call the constellation by a human name, they could not actually draw a human figure, so in their representations of Virgo the 'innocent maiden' turns out as a sheaf of wheat, with Spica, naturally enough, as its ripest and brightest ear. Not surprisingly the name of the constellation has also been associated with many young women beside those mentioned here, including Erigone (who hanged herself in grief on the death of her father), Irene (the goddess of peace), Diana (goddess of the hunt), Minerva (the Roman equivalent of Athene), Isis (the main goddess of ancient Egypt), Ishtar (the Babylonian goddess of love who was the 'prototype' of Aphrodite) and the Virgin Mary. Whoever else Virgo was, she was consistently a symbolic representation of innocence and virtue. Despite the admirable (if idealistic) personification of the stars here, it has to be said that it is hard to make out any obvious resemblance to a human figure, virginal or otherwise.

Volans (constellation)
This is the shortened form of the original constellation name of Piscis Volans, The Flying Fish, introduced in 1603 by the German

astronomer Johann Bayer. The name is typical of his fairly exotic creatures, and the kite-shaped outline of the main stars here could be taken for a flying fish of some kind, with Delta, Gamma and Zeta Volantis marking the front of its head, and Alpha Volantis the tip of its tail. The name also happens to fit in well with those of its neighbouring constellations of *Carina* and (better still) *Dorado*. The latter was also introduced by Bayer, but not the former, which was a creation of Lacaille in the mid-eighteenth century.

Vulcan (non-existent planet)
This is a real 'ghost name', belonging to a planet that does not exist. In the nineteenth century the French astronomer Urbain Le Verrier reasoned that there could be a planet between *Mercury* and the Sun, making it the nearest to the latter in the Solar System. It was named Vulcan, after the Roman god of fire, and one must admit that this would have been a suitable name for what would certainly have been the 'hottest' planet, nearest to the flames of the Sun. However, its existence was never proved, so Vulcan remained simply a desirable name. Le Verrier had much more success when he set out to prove the existence of a trans-Uranian planet (*Neptune*) than an intra-Mercurial one.

Vulpecula (constellation)
The name of this constellation, introduced by the German astronomer Hevelius in the late seventeenth century, was in full Vulpecula cum Ansere, The Fox with the Goose (or to be more precise, 'The Little Fox with the Goose'). We are fortunate enough to have a detailed explanation for his particular choice of name, in the following extract from the writings of the nineteenth-century English astronomer William Henry Smyth: 'Hevelius [wished] to occupy a space between the Arrow and the Swan, where the Via Lactea divided into two branches. For this purpose he ransacked the *informes* of this bifurcation, and was so satisfied with the result, that the effigies figure in the elaborate print of his offerings to Urania. He selected it on account of the Eagle, Cerberus and Vultur Cadens. "I wished," said he, "to place a fox and a goose in the space of the sky well fitted to it; because such an animal is very cunning, voracious and fierce. Aquila and Vultur are of the same nature, rapacious and greedy." ' ('The Arrow' is *Sagitta*, 'the Swan', *Cygnus*, and 'the Via Lactea', the *Milky Way*. By 'ransacking the *informes*' he means that he adopted unnamed figures from the pictorial maps of the Milky Way. 'Urania' is the

Roman muse of astronomy, so a hoped-for inspirer of new names! 'The Eagle' is *Aquila*, and 'Vultur Cadens' is another name of Cygnus, comparable to the 'Swooping Eagle' of *Lyra*, seen in the name of *Vega*.) But if there was a goose once, the bird has now flown, and we are left simply with the 'cunning little vixen'.

Water Carrier (Water Bearer), The see **Aquarius**

Water Snake, The see **Hydra**

Wezen (star Delta Canis Majoris)
Rather unexpectedly the name turns out to mean 'the weight', from Arabic Al Wazn. If this is the correct interpretation, one must assume that the star was so called because it appeared to rise slowly and 'weightily' above the horizon. It can hardly have anything to do with the depiction of *Canis Major* itself, where it is located near the dog's hind legs. Or is it supposed to have some function with regard to his tail? Perhaps the tail itself is thought of as a 'weight' that the dog has to carry, much as a racehorse's jockey is spoken of as its 'weight'.

Whirlpool Galaxy (in *Canes Venatici*)
The galaxy, catalogued as NGC 5194 (or M 51), has a nickname which describes its swirling, spiral appearance when seen through a fairly powerful telescope.

Wild Duck Cluster (in *Scutum*)
Officially known as NGC 6705 (or M 11), the spectacular and dense star cluster is rather more adventurously nicknamed from its fanlike appearance, which suggests a host of wild ducks flying out in all directions from a central point. The name was given by Admiral W. H. Smyth (see *Mare Smythii*).

Winged Horse, The see **Pegasus**

Wolf, The see **Lupus**

Yed Posterior (star Epsilon Ophiuchi)
The half-Arabic, half-Latin name means literally 'hand behind', that is, the star that marks the second of the pair of stars which depict the hand of *Ophiuchus* as he holds the head of the snake. Compare *Yed Prior*.

Yed Prior (star Delta Ophiuchi)
This name is the 'other half' of the Serpent Holder's hand described in the entry for *Yed Posterior*, and so means 'hand before', because it lies to the north of its fellow.

Zosma (star Delta Leonis)
Many names beginning with 'z', such as the ones below, are Arabic in origin. This one, however, is Greek, and derives from *zōma*, 'girdle', because this is what it is supposed to mark on the figure of *Leo*, The Lion. It is actually on the animal's back, however, near its tail, so that the name may either be a corruption after all, or else have originally applied to some other star. The Arabic name for this star, more logically, is Al Ṭhahr al Asad, 'the back of the lion'.

Zubenelakrab (star Gamma Librae)
The Arabic name is usually said to mean 'claw of the scorpion', and to refer, like *Zubenelgenubi* and *Zubeneschamali*, to the earlier representation of the constellation of *Libra* as an extension of neighbouring *Scorpius*, with the names corresponding to the subsequent Greek 'Chelē', 'claw'. But modern scholarship sees the origin of the main part of the name in much earlier Sumerian ZIB-BA AN-NA, 'balance of heaven', to which has been added Arabic Al Akrab, 'the scorpion'. This certainly accords well with the name of the star's constellation as a whole. The present name should really apply to Alpha Librae, that is, to *Zubenelgenubi*, so that the latter name belongs properly to Sigma Librae (which was originally Gamma Scorpii).

Zubenelgenubi (star Alpha Librae)
As for *Zubenelakrab*, the derivation was long said to lie in the Arabic for 'the southern claw', to which subsequent Greek 'Chelē notios' corresponded. But it is probably better to take the Sumerian meaning of 'balance of heaven' (see *Zubenelakrab*), with Arabic Al Janūbiyyah, 'the south', added. In any case, as it stands this star is not the southern claw of neighbouring *Scorpius*, and the present name really belongs to Sigma Librae, as mentioned above. Compare *Zubeneschamali*.

Zubeneschamali (star Beta Librae)
As for *Zubenelakrab* and *Zubenelgenubi*, the first part of the name is usually said to derive from Arabic Al Zubān, 'the claw'. It is more convincing, however, to adopt the new interpretation given for these names and trace the origin back to Sumerian, meaning 'balance of heaven' (see *Zubenelakrab*). To this, Arabic Al Shamāliyyah, 'the north', has been subsequently added. The present name was formerly applied to Alpha Librae, now called *Zubenelgenubi*. The confusion in the recent nomenclature of these three

stars is due to the fact that the Greeks regarded The Scales as an extension of The Scorpion, as mentioned.

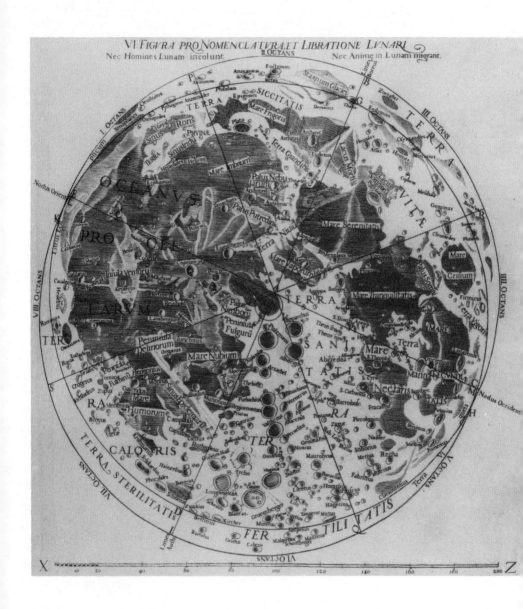

The near side of the moon from *Almagestum Novum*,
Vol. I, Riccioli. (*Courtesy Royal Astronomical Society*)

APPENDIX I: NAMES OF CRATERS ON THE MOON

The following list aims to identify the persons, real or mythological, who have given their names to craters on the Near Side and Far Side of the Moon, and the names themselves are based closely on the listing of over 1000 craters (together with some other features) included as a 'Crater Index' in Patrick Moore and Garry Hunt's *The Atlas of the Solar System* (see Bibliography).

Each crater is given its position (selenographical coordinates) on the Moon's surface, and is followed by a brief 'identity statement' regarding the person who gave it its name, including his or her nationality, profession or 'relevant activity', and years of birth and (where it has occurred) death. A word should be said about each of these components.

The position of the crater or other feature is given by two co-ordinates, the first being its latitude (in degrees north or south of the Moon's 'equator'), and the second its longitude (in degrees east or west of the Moon's meridian, which runs as a bearing of 0° from the north lunar pole to the south through the appropriately named Sinus Medii). The coordinates are the only means the reader will have for determining whether the crater is on the Near Side or the Far Side of the Moon. The actual site of the crater may in itself seem not always relevant, yet the names on the Far Side are, of course, the most recent and therefore among the more interesting, and at the same time some names can be found arranged in relevant groupings, which is also significant. But as a general principle, all names with a bearing greater than 90° of longitude will be on the Far Side of the Moon, so this is the significant coordinate here. Thus the crater named Abbe (the first in the list) can be seen to be on the Far Side, with its longitude of 175°E, but Abel (the second in the list) is on the Near Side, with its longitude of less than 90°. From this information it broadly follows that a Far Side name will be one given after 1961 (of which more below), while a Near Side name will usually have been allotted before this date.

The name of the person is usually given in conventional form, with first name(s) or initials, and sometimes professional or honorary designation, such as 'Sir' or 'Comte'. Many of the names of historic personages, especially those who lived several hundred years ago, are assigned to a crater in the Latin form of the name, and where this happens the indigenous (or better-known) version of the name is given in brackets. For example the twelfth-century Arab astronomer called Alpetragius was known in his own language as al-Betrugi, and the sixteenth-century Italian astronomer Gauricus was called Luca Gaurico in his native tongue. (The fashion for Latin names was common among Renaissance scholars, with names either 'Latinised' or even translated: in astronomy and geography we have two good examples in Copernicus and Mercator, whose true names were Kopernik and Kremer, the latter meaning 'merchant'. Such names would not necessarily have been devised specially for lunar nomenclature, but would have already been in use.)

Where the name of an individual coincides in common use with the name allocated to the particular crater, it is not repeated, as is the case with many names of the 'ancients', such as the Greek philosophers Anaxagoras and Plato.

The most common profession or 'relevant activity' stated is that of 'astronomer', of course, but the designation can have a greater implication both historically and notionally. The ancient Greek and Roman astronomers were often what we would now call 'astrologers'; conversely, most astronomers today work in an academic field that not only centres on astronomy but also embraces various other relevant disciplines, such as mathematics and physics. To see 'astronomer' by a fairly modern name means more than it implies, therefore. For example the American astronomer Joel Stebbins could more accurately be described as a stellar photometrist, in which narrower field he was a pioneer, and the German astronomer Julius Schmidt was in particular a lunar cartographer, or 'moon map-maker'. The reverse can equally apply, so that 'physicist' can imply an involvement with astronomy, and 'mathematician' can embrace both astronomy and physics. In a sense, though, all these descriptive designations are secondary to the main purpose of the Appendix, which is to identify the particular individual after whom a crater is named.

Even so, the 'job descriptions' *are* of interest, especially when they break the conventional mould. Here are the names of travellers and explorers, for instance, such as Amundsen and Scott, Nansen and Peary. Moreover, as observant readers will spot by examining their particular coordinates, the first two craters are appropriately located near the north pole of the Moon, while the latter pair are near the south pole, with each polar explorer thus commemorated in the cartographical area associated with him. Then there are the two novelists, Jules Verne and H. G. Wells, each of whom has his own crater. In this scientific world, the names of people prominent in the arts are not common, but Verne and Wells are here. The Moon played a prominent part in the imaginative works of each: Verne wrote *From the Earth to the Moon* and Wells, *The First Men in the Moon*. These Victorian fantasies have now, as we know, become realities.

Back in the more general world of science, it will be seen that many of the names commemorated on the Moon either are those of inventors, or have even become standard words in their own right. We thus find Bunsen of the Bunsen burner, Davy of the Davy lamp, Dewar of the Dewar flask, Doppler of the Doppler effect, Faraday of the farad (the unit of capacitance), Fermi of the fermi (the unit of length), Galvani of galvanism, Geiger of the Geiger counter, Hertz of the hertz (the unit of frequency), Maxwell of the maxwell (the unit of magnetic flux), Morse of the Morse code, Nobel of the Nobel prizes, Ohm of the ohm (the unit of electrical resistance), Pasteur of pasteurised milk, Tesla of the tesla (the unit of magnetic flux), Volta of the volt (the unit of electrical potential), Watt of the watt (the unit of power) and many others. Some of these, although long well known, were first added as names on the Far Side (for example Hertz, Maxwell, Pasteur), which makes one wonder how they could have been overlooked before.

But whatever the link, scientific or otherwise, there must be some connection with astronomy or in particular with the Moon for the name to have been given at all. This must therefore apply for names without any immediately obvious connection. How about Chaucer, for example? What do the *Canterbury Tales* have to do with the Moon? The answer does not lie in this, his most famous work, but in his lesser known *Treatise on the Astrolabe*, written for 'little Lewis', his son. Readers may enjoy tracking down other lunar links for similar names.

Many of the more recent names are those of 'rocket pioneers', of spaceflight experts and scientists, and some are of actual astronauts or cosmonauts. We shall return to them shortly.

171

Appendix I: Names of Craters on the Moon

The nationality of the name-giver provides a sort of visual guide to the history of astronomy, with those nationalities dominant where the greatest steps to the stars have been taken. On nearly every page one sees 'German', 'American', 'English', 'Scottish', for example. In fact, a 'nationality count' of the total 1000-plus names shows that just over 18 per cent of all names are German, nearly 15 per cent are American, and just over 13 per cent are British, including 'Scottish' but not 'Irish'. (The Scots are particularly well represented in the British contingent, in fact, as a fitting tribute to that country's significant contribution to the sciences and to astronomy in particular.) After these three leaders, names can be found in the following order of frequency: French (around 11 per cent), Russian (some 10 per cent), then Greek and Italian (about 6 per cent each), Dutch, Austrian, Roman (Latin), Danish, Swedish, Swiss (long live Paracelsus!) and Arabian. Bottom of the lunar league come the Belgian, Spanish, Polish, Hungarian, Norwegian, Czech, Irish, Japanese, Canadian and Indian names. A few countries, such as Finland, Yugoslavia, Portugal and Australia, are represented by fewer than five names, and some, such as Egypt or Turkey, are not represented at all (although the list is not itself exhaustive).

'Greek' here means almost exclusively 'Ancient Greece', just as 'Roman' relates to classical Rome and 'Arab' implies a similar historicity. 'Russian' can be either pre-1917, before the Revolution, and so be 'Russian' proper in the listing, or post-1917, when it is given as 'Soviet'. (An astronomer who was born even some time before the Revolution but whose work was carried out to a significant extent after it will still be designated as 'Soviet', as he worked in the Soviet era.)

The complexities of a particular person's national background are not gone into here; many eminent astronomers were born in one country but carried out their life's work in another. Only a general guide to such a multinational background can be given in the brief entries here.

Dates of birth and death are given as precisely as possible, but the further back one goes in history, the more uncertain they become, so that for some of the astronomers or scholars only a century can be given. A query ('?') appears before those years that remain uncertain.

It will be noticed that, almost without exception, the names are of people who are no longer living. The exception can be found for a few living American and Soviet astronauts and cosmonauts.

172

This brings us right into the present, and will serve as a timely juncture for the more recent history of Moon-naming to be considered briefly here.

The early history of lunar nomenclature has already been outlined in the Introduction (pp. 9–14), where the origin of the names of the major lunar features is given, such as the 'oceans', 'seas' and 'mountains' (*oceani*, *maria* and *montes*). After the initial namings by Langren, Hevelius, Riccioli and Grimaldi, and the subsequent additions and rearrangements by Schröter and others, things remained in a rather confused and random state, with various selenographers giving and altering names independently of one another, until in the twentieth century, some 300 years after the original namings, two leading astronomers determined to bring some order out of the chaos, and to draw up a unified nomenclature for the features on the Moon. These were Miss Mary Blagg, the English astronomer (more precisely selenographer), who is now commemorated by a crater of her own on the Moon, and Dr Karl Müller, of Vienna, a similar selenographical expert. The results of their joint work were a map of the Moon and an accompanying catalogue, prepared in 1926 and officially approved by the International Astronomical Union in 1932. It was then published in book form (see Blagg and Müller, *Named Lunar Formations*, in the Bibliography).

Further amendments and alterations were made to their recommendations in the 1960s, with the rapid expansion of a more sophisticated type of selenography, and it was in this same decade that, for the first time, names began to be given to features on the Far Side of the Moon. This was a direct consequence of the historic flight round the Moon by the Soviet spaceprobe Luna 3 in October 1959, when the first photographs of the Far Side were taken and when features could be identified and named. Then, exactly ten years later, the first landing on the Moon took place, when in July 1969 the American lunar craft Apollo 11 came to rest on the surface and Neil Armstrong stepped out onto it. The crater Apollo was named as a commemoration of this unique event.

Most names of lunar features today are those recognised by the International Astronomical Union at its 14th General Assembly in 1970, when 513 new names were put forward for acceptance. The majority of these were for features on the newly charted Far Side, although some additions and adjustments were also made in the nomenclature of the Near Side. For example the name of Porter was at first proposed for a crater on the Far Side, but on the suggestion of the British astronomer Patrick Moore was given to a crater (inside Clavius) on the Near Side.

Originally it had been proposed to assign the new names to the Far Side in roughly alphabetical order, from the north pole to the south, but this plan was objected to on aesthetic grounds by several cartographers, and so was not implemented.

How do Far Side names differ from those of the Near Side? Clearly here was a golden opportunity to commemorate leading scientists and astronomers of the twentieth century, and such names are certainly conspicuous here. But many nineteenth-century names were also added, and even ancient ones, such as that of Hippocrates, the Greek physician of the fifth and fourth centuries BC. Even one or two mythological names were assigned to Far Side craters, among them Daedalus and his son Icarus. (Their coordinates, respectively 6S, 180 and 6S, 173W, show them to be keeping close company.)

In a few instances, names proposed for craters on the Far Side could not be practically used because they closely resembled names on the Near Side, and could thus cause confusion. For example, one obvious candidate was the name Rutherford, for the British physicist and Nobel prizewinner. But this could not be assigned as the American astronomer Rutherfurd was already on the Near Side. Similarly the name of the British nuclear physicist Max Born could not be added, as the Danish physicist Niels Bohr was already on the Near Side.

It is in the names of American and Soviet astronauts and cosmonauts, however, that one of the most distinctive innovations was made. They, after all, had performed a unique feat: they had not only travelled in space, but also in the case of the Americans had actually walked on the surface of a celestial body, the Moon. However great the discoveries and achievements of astronomers and selenographers had hitherto been, they had nevertheless remained firmly Earth-bound, and had not physically entered the medium and made contact with the subject of their study.

Both American and Soviet delegations to the 1970 conference thus proposed that the names of six living astronauts and six living cosmonauts should be added to the names of deceased astronauts and cosmonauts that had already been commemorated on the surface. This plan broke with tradition, and at first met with some opposition. Eventually the proposal was agreed and so the names of the three American astronauts of Apollo 8, Anders, Borman and Lovell (who were the first to fly round the Moon, in 1968), were given to craters near the crater Apollo, on the Far Side, where they joined their deceased colleagues, Chaffee, Grissom and White, while the three-man crew of Apollo 11, Aldrin, Armstrong and Collins, had their names assigned to three craters near their craft's landing point on the Mare Tranquillitatis. Similarly the names of the six living Russian cosmonauts, Feoktistov, Leonov, Nikolaev, Shatalov, Tereshkova and Titov, were awarded to craters near the Mare Moscoviense ('Sea of Moscow') where their former colleagues, Komarov, Belyayev, Volkov, Dobrovolsky and Patsayev, were also commemorated. Gagarin, although likewise a cosmonaut – and moreover that of the first pioneering manned space flight in 1961 – has his own crater some distance south of the Mare Moscoviense. (As already explained, the listing below is not fully comprehensive, so that not all these names will be found there.)

At the same time, it was proposed that the actual landing site of Apollo 11 should be officially named Statio Tranquillitatis ('Tranquillity Base'), and this will now be found, marked with a small 'x', on most lunar maps of any detail. In the same way, the site where the Soviet spacecraft Luna 2 impacted in 1959 was designated Sinus Lunicus ('Luna Bay'), and the landing site of Luna 9 (in the Oceanus Procellarum, 1966) was named Planitia Descensus ('Landing Plain').

It is perhaps also worth mentioning that some names already given to features on the Far Side were now deleted, one of them being the apparent mountain chain known as Montes Sovietici ('Soviet Mountains'). When this 'range' was found to be not a range at all, but simply a bright ray, the name was removed from the lunar map, as were the names of four other so-called 'ranges'.

It will be noticed that some craters are named after two or even three people, often members of the same family, although not always so. This is a fitting and convenient way to commemorate or honour astronomical namesakes, and any family ties, when known, are indicated in the list below. But in one or two instances the converse can be found, when different members of a family have their own individual crater. This can cause difficulties, as it is likely they will share a common surname. An example of this occurred with the famous Curie family, who included the French chemist Pierre Curie, his wife Marie Curie, and their physicist son-in-law Frédéric Joliot-Curie. (The latter was originally Frédéric Joliot, but took the name Joliot-Curie on marrying Pierre and Marie Curie's daughter Irene.) Frédéric Joliot-Curie already had a crater named after him, as (logically enough) Joliot-Curie. When the names of Pierre and Marie Curie were selected for addition, however, three distinctive names were required for the three separate craters. The Commission resolved the matter by renaming Joliot-Curie as Joliot, by calling Pierre Curie's crater Curie, and by giving Marie Curie's crater her maiden name, Skłodowska. (Although now generally known as Marie Curie, she was Polish by birth, and before her marriage was Marya Skłodowska.)

Meanwhile there are still unnamed craters and features available for any future Moon names, and it should not be thought that the naming of lunar features is now complete. (Naming can similarly be continued on other celestial objects, and in 1986, for example, Soviet cartographers decided to name two craters on Venus after the two American women who lost their lives in the Challenger disaster earlier that year, the teacher Christa McAuliffe and the astronaut Judy Resnik. Ultimately, however, all such name proposals must be officially authorised and approved by the International Astronomical Union.)

Since 1970 a number of craters with letter designations based on the name of a nearby crater (in the manner explained in the Introduction, see p. 14) have been given names of their own. For example Macrobius A, Macrobius B, Macrobius D and Macrobius L, all designated after nearby Macrobius, are now respectively known as Carmichael, Hill, Fredholm and Esclangon. These are all minor names, however, and are not included in the list below.

The Moon has not only its official names but also a small range of nicknames. These were devised, more or less spontaneously, by members of various American Apollo missions for local features that needed identifying and that were not already officially named. But although they were nicknames, their use was recognised by mission control and was indeed recorded in order to facilitate the different operations of the missions. For example, names used for local features by the crew of Apollo 12 (Conrad, Bean and Gordon) when they landed in the Oceanus Procellarum in 1969 included Bench, Block, Crescent, Halo, Head, Snowman and Surveyor. Apollo 14, which landed with Shepard, Mitchell and Roosa on board in 1971, used the names Cone, Doublet, Flank, Old Nameless and Weird, while Apollo 15 (landed six months later with Scott, Irwin and Worden as crew) made use of names such as Dune, Earthlight, Elbow, Index, Spur and St George. The following year, Apollo 16 landed in the Descartes area with Young, Duke and Mattingly as crew, and devised fairly homely names for different craters and clusters, such as Baby Ray, Flag, Gator, Plum, Spook, Spot, Stubby and Wreck. Other features were called by them Smoky Mountains and Stone Mountains.

These were put in the shade onomastically by the crew of Apollo 17, however, and when Cernan, Schmitt and Evans landed near the Taurus Montes in 1972 they deployed much more literary-sounding names. Among them were craters and clusters called Brontë, Camelot, Horatio, Powell, Shakespeare, Sherlock and Trident, with other features called Family Mountain, Sculptured Hills, Tortilla Flat (*pace* Steinbeck) and Wessex Cleft (acknowledgments to Hardy). All the same, they did not go quite overboard here in their literary euphoria, and found space for a rock cluster called Shorty and a more homely Bear Mountain, redolent of their native Rockies.

Apollo 17 was the last manned landing on the Moon to date, so that we have yet to see how subsequent lunar nickname creation will develop; presumably it will be much more purposeful and self-conscious.

But what makes these sometimes frivolous nicknames unique is that they are the names of physical objects visited by real people, and so are exactly the same as the many ancient terrestrial place-names with which we are much more familiar. Many of those began their lives as descriptive nicknames, too, even though the original word may have become distorted with the passage of time so that it is no longer recognisable.

It would have been interesting to see what nicknames the Russians would have devised had they also landed on the Moon. They would probably have been very much on the same lines, one imagines, with due allowance for literary and other national and political propensities. (The American nicknames do not of course appear on standard maps of the Moon, but they do feature on very large-scale maps and photographs of the respective Apollo landing sites.)

Finally, it should be stated that in 1982 there were a total of 1395 named craters on the Moon (796 on the Near Side and 599 on the Far Side), while there were over 7000 craters simply designated by letter. This means that lunar namers still have plenty of scope for the exploitation of their imagination, intelligence and general inventiveness!

Names in the list below that have their own entries in the Dictionary (mainly features that are not craters) are printed in *italics*, while the list as a whole is a virtual 'Who's Who' of astronomers mentioned elsewhere in the book, including those additionally featuring in the names of minor planets (see Appendix II).

Abbe (58S, 175E): Ernst Abbe, German physicist (1840–1905)
Abel (36S, 85E): Niels Hendrik Abel, Norwegian mathematician (1802–29)
Abenezra (21S, 12E): Abraham Abenezra, Spanish-Jewish scholar (?1093–1167)
Abulfeda (14S, 14E): Ismael Abulfeda, Arab geographer and historian (1273–1331)
Abul Wáfa (2N, 117E): Persian astronomer and mathematician (940–98)
Adams (32S, 69E) (1) John Couch Adams, English astronomer (1819–92);
            (2) Charles H. Adams, American astronomer (1868–1951);
            (3) Walter S. Adams, American astronomer (1876–1956)
Aestuum, Sinus (12N, 9W)
Agatharchides (20S, 31W): Greek historian and geographer, second century BC
Agrippa (4N, 11E): Greek astronomer, first century AD
Airy (18S, 6E): George Biddell Airy, English astronomer (1801–92)
Aitken (17S, 173E): Robert G. Aitken, American astronomer (1864–1951)
Albategnius (12S, 4E): (al-Battani) Arab astronomer (?850–929)
Al Biruni (18N, 93E): Arab scholar and author (973–1048)
Alden (24S, 111E): Harold L. Alden, American astronomer (1890–1964)
Alekhin (68S, 130W): Nikolai P. Alekhin, Soviet rocket design engineer (1913–63)
Alfraganus (6S, 19E): (al-Farghani) Arab astronomer, ninth century AD
Alhazen (18N, 70E): (al-Hasan) Arab astronomer (?965–?1039)
Aliacensis (31S, 5E): (Pierre d'Ailly) French theologian (1350–1420)
Almanon (17S, 15E): (al-Mamun) Arab philosopher and astronomer (786–833)
Alpes, Vallis (49N, 2E): 'Alpine valley' cutting through Montes Alpes, 'Alps'

Alpetragius (16S, 4W): (al-Betrugi) Arab astronomer, twelfth
century
Alphonsus (13S, 3W): (Alfonso X, 'El Sabio') King of Castile
(?1226–84)
Altai, Rupes (24S, 22E): 'Altai rift', also known as 'Altai Scarp'
Alter (19N, 108W): Dinsmore Alter, American astronomer
(1888–1968)
Amici (10S, 172W): Giovanni B. Amici, Italian astronomer
(?1786–?1863)
Amundsen (83S, 103W): Roald E. Amundsen, Norwegian polar
explorer (1872–1928)
Anaxagoras (75N, 10W): Greek philosopher (?500–428)
Anaximander (66N, 48W): Greek astronomer (612–547)
Anaximenes (75N, 45W): Greek philosopher, sixth century BC
Anděl (10S, 13E): Karel Anděl, Czech astronomer (1884–1947)
Anders (42S, 144W): William A. Anders, American astronaut
(born 1933)
Anderson (16N, 171E): John A. Anderson, American astronomer
(1876–1959)
Ångström (30N, 42W): Anders J. Ångström, Swedish astronomer
and physicist (1814–74)
Anguis, Mare (23N, 69E): 'Serpent Sea', narrow valley named by
Franz (see below)
Ansgarius (14S, 82E): (St Anschar) Frankish missionary, (801–865)
Antoniadi (69S, 173W): Eugenios Antoniadi, Greek astronomer
(1870–1944)
Apenninus, Montes (20N, 2W): see *Apennines*
Apianus (27S, 8E): (Peter Bienewitz) German astronomer
(?1501–52)
Apollo (37S, 153W): for American spacecraft Apollo 11, landing
on Moon in 1969
Apollonius (5N, 61E): Greek mathematician, third century BC
Appleton (37N, 158E): Sir Edward V. Appleton, English physicist
(1892–1965)
Arago (6N, 21E): François Arago, French scientist (1786–1853)
Aratus (24N, 5E): Greek physician and poet (?315–?245)
Archimedes (30N, 4W): Greek mathematician (?287–212)
Archytas (59N, 5E): Greek philosopher and scientist, fourth
century BC
Argelander (17S, 6E): Friedrich Argelander, German astronomer
(1799–1875)

Ariadaeus, Rima (7N, 13E): 'Ariadaeus' rill' (Arrhidaeus, King of Macedonia, put to death 317 BC)

Aristarchus (24N, 48W): Greek astronomer, third century BC

Aristillus (34N, 1E): Greek astronomer, third century BC

Aristoteles (50N, 18E): (Aristotle) Greek philosopher (384–322)

Arnold (67N, 38E): Christoph Arnold, German amateur astronomer (1650–95)

Arrhenius (55S, 91W): Svante A. Arrhenius, Swedish astronomer (1859–1927)

Artamonov (26N, 104E): Nikolai N. Artamonov, Soviet rocket engineer (1906–65)

Artem'ev (10N, 145W): Vladimir A. Artem'ev, Soviet rocket designer (1885–1962)

Arzachel (18S, 2W): (Abraham Elzarakeel) Spanish-Arab astronomer, eleventh century

Asclepi (55S, 26E): Giuseppe Asclepi, Italian astronomer and physicist (1706–76)

Aston (32N, 88W): Francis William Aston, English physicist (1877–1945)

Atlas (47N, 44E): Titan in Greek mythology

Australe, Mare (50S, 80E)

Autolycus (31N, 1E): Greek mathematician and astronomer, fourth century BC

Auwers (15N, 17E): Arthur von Auwers, German astronomer (1838–1915)

Auzout (10N, 64E): Adrien Auzout, French astronomer (1622–91)

Avicenna (40N, 97W): (ibn-Sina) Arab physician and philosopher (980–1037)

Avogadro (64N, 165E): Amedeo Avogadro, Italian chemist (1776–1856)

Azophi (22S, 13E): (Al Sufi) Arab astronomer (903–86)

Baade (47S, 83W): Walter Baade, German-born American astronomer (1893–1960)

Babbage (58N, 52W): Charles Babbage, English mathematician (1792–1871)

Babcock (4N, 94E): Harold D. Babcock, American physicist (1882–1968)

Backlund (16S, 103E): Oscar A. Backlund, Swedish-born Russian astronomer (1846–1916)

Baco (51S, 19E): (Roger Bacon) English philosopher and scientist (?1214–94)

Baillaud (75N, 40E): Benjamin Baillaud, French astronomer (1848–1934)

Bailly (66S, 65E): Jean Sylvain Bailly, French astronomer (1736–93)

Baily (50N, 31E): Francis Baily, English astronomer (1774–1844) (see *Baily's Beads*)

Balboa (20N, 84W): Nuñez de Balboa, Spanish explorer (1475–1519)

Baldet (54S, 151W): François Baldet, French astronomer (1885–1967)

Ball (36S, 8W): William Ball, English astronomer (died 1690)

Balmer (20S, 70E): Johann Jakob Balmer, German mathematician (1825–98)

Barbier (24S, 158E): Daniel Barbier, French astronomer (1907–65)

Barker (42S, 8W): Lewellys F. Barker, American physicist (1867–1943)

Barnard (32S, 86E): Edward E. Barnard, American astronomer (1857–1923) (see *Barnard's Star*)

Barocius (45S, 17E): (Francesco Barozzi) Italian mathematician (died ?1570)

Barringer (29S, 151W): Daniel M. Barringer, American geologist (1860–1929)

Barrow (73N, 10E): Isaac Barrow, English mathematician (1630–77)

Bayer (51S, 35W): Johann Bayer, German astronomer (1572–1625)

Beaumont (18S, 29E): Élie de Beaumont, French geologist (1798–1874)

Becquerel (41N, 129E): Antoine H. Becquerel, French physicist (1852–1908)

Bečvář (2S, 125E): Antonín Bečvář, Czech astronomer (1901–65)

Beer (27N, 9W): Wilhelm Beer, German astronomer (1797–1850)

Behaim (16S, 71E): Martin Behaim, German navigator (?1459–?1507)

Beijerinck (13S, 152E): Martin Beyerinck, Dutch microbiologist (1851–1931)

Belkovich (60S, 92E): Igor V. Belkovich, Soviet astronomer (1904–49)

Bell (22N, 96W): Alexander Graham Bell, Scottish physicist (1847–1922)

Bellingshausen (61S, 164W): Faddei F. Bellingshausen, Russian explorer (1778–1852)

Belopolsky (18S, 128W): Aristarkh A. Belopolsky, Soviet astronomer (1854–1934)

Belyayev (23N, 143E): Pavel I. Belyayev, Soviet cosmonaut (1925–70)

Bergstrand (19S, 176E): Carl Ö.E. Bergstrand, Swedish astronomer (1873–1948)

Berkner (25N, 105W): Lloyd V. Berkner, American physicist (1905–67)

Berlage (64S, 164W): Hendrik P. Berlage, Dutch geophysicist (1896–1968)

Bernouilli (34N, 60E) (1) Jacques Bernouilli, Swiss mathematician (1654–1705);
(2) Jean Bernouilli, Swiss mathematician (1667–1748), brothers

Berosus (33N, 70E): Babylonian priest and astronomer, third century BC

Berzelius (37N, 51E): Jöns Jakob Berzelius, Swedish chemist (1779–1848)

Bessarion (15N, 37W): Johannes Bessarion, Greek scholar and prelate, fifteenth century

Bessel (22N, 18E): Friedrich Wilhelm Bessel, German astronomer (1784–1846)

Bettinus (63S, 45W): (Mario Bettini) Italian scholar (1582–1657)

Bhabha (56S, 165W): Homi J. Bhabha, Indian physicist (1909–66)

Bianchini (49N, 34W): Francesco Bianchini, Italian astronomer (1662–1729)

Biela (55S, 52E): Wilhelm von Biela, Austrian astronomer (1782–1856) (see *Biela's Comet*)

Billy (14S, 50W): Jacques de Billy, French astronomer (1602–79)

Biot (23S, 51E): Jean Baptiste Biot, French astronomer (1774–1862)

Birkeland (30S, 174E): Olaf K. Birkeland, Norwegian physicist (1867–1917)

Birkhoff (59N, 148W): George D. Birkhoff, American mathematician (1884–1944)

Birmingham (64N, 10W): John Birmingham, Irish astronomer (1829–84)

Birt (22S, 9W): William Radcliffe Birt, English astronomer (1804–81)

Bjerknes (38S, 113E): Vilhelm F.K. Bjerknes, Norwegian physicist (1862–1951)

Blagg (1N, 2E): Mary Adela Blagg, English astronomer (1858–1944)

Blancanus (64S, 21W): (Giuseppe Biancani) Italian astronomer (1566–1624)

Blanchinus (25S, 3E): (Giovanni Bianchini) Italian astronomer (died ?1458)

Blazhko (31N, 148W): Sergei N. Blazhko, Soviet astronomer (1870–1956)

Bobone (29N, 131W): Jorge Bobone, Argentine astronomer (1901–58)

Bode (7N, 2W): Johann Elert Bode, German astronomer (1747–1826)

Boguslawsky (75S, 45E): Ludwig von Boguslawsky, German astronomer (1789–1851)

Bohnenberger (16S, 40E): Johann von Bohnenberger, German astronomer (1765–1831)

Bohr (13N, 86W): Niels Bohr, Danish physicist (1885–1962)

Boltzmann (78S, 99E): Ludwig Boltzmann, Austrian physicist (1844–1906)

Bolyai (34S, 125E): Janos Bolyai, Hungarian mathematician (1802–60)

Bond, G. (32N, 36E): George P. Bond, American astronomer (1826–65)

Bond, W. (64N, 3E): William C. Bond, American astronomer (1789–1859)

Bonpland (8S, 17W): Aimé Bonpland, French naturalist (1773–99)

Borda (25S, 47E): Jean Charles de Borda, French astronomer (1733–99)

Borman (37S, 142W): Frank Borman, American astronaut (born 1928)

Boscovich (10N, 11E): Ruggiero Giuseppe Boscovich, Italian astronomer (1711–87)

Bose (54S, 170W): Jagadis C. Bose, Indian physicist (1858–1937)

Boss (46N, 90E): Lewis Boss, American astronomer (1846–1912)

Bouguer (52N, 36W): Pierre Bouguer, French mathematician (1698–1758)

Boussingault (70S, 50E): Jean-Baptiste Boussingault, French chemist (1802–87)

Boyle (54S, 178E): Robert Boyle, Irish physicist and chemist (1627–91)

Bragg (42N, 103W): Sir William Henry Bragg, English physicist (1862–1942)

Brashear (74S, 172W): John A. Brashear, American lens-maker (1840–1920)

Brayley (21N, 37W): Edward William Brayley, English science writer (1802–70)

Bredikhin (17N, 158W): Fyodor A. Bredikhin, Russian astronomer (1831–1904)

Breislak (48S, 18E): Scipione Breislak, Italian geologist (1748–1826)

Brenner (39S, 39E): Leo Brenner (real name Spiridon Gopcevič), Austrian amateur astronomer (1855–1928)

Brianchon (77N, 90W): Charles Julien Brianchon, French mathematician (1785–1864)

Bridgman (44N, 137E): Percy W. Bridgman, American physicist (1882–1961)

Briggs (26N, 69W): Henry Briggs, English mathematician (?1556–1631)

Brisbane (50S, 65E): Sir Thomas M. Brisbane, Scottish astronomer (1773–1860)

Brouwer (36S, 125W)   (1) Dirk Brouwer, American astronomer (1902–66);
                      (2) Luitzen E. J. Brouwer, Dutch mathematician (1881–1966)

Bruce (1N, 0): Catherine W. Bruce, American art and science patron (1816–1900)

Brunner (10S, 91E): William O. Brunner, Swiss astronomer (1878–1958)

Buch (39S, 18E): Christian Leopold von Buch, German geologist (1774–1853)

Buffon (41S, 134W): Comte George de Buffon, French naturalist (1707–88)

Buisson (1S, 113E): Henri Buisson, French astronomer (1873–1944)

Bullialdus (21S, 22W): (Ismael Boulliaud) French astronomer (1605–94)

Bunsen (41N, 85W): Robert Wilhelm Bunsen, German chemist (1811–99)

Burckhardt (31N, 57E): Johan Karl Burckhardt, German astronomer (1773–1825)

Bürg (45N, 28E): Johann Tobias Bürg, Austrian astronomer (1766–1834)

Burnham (14S, 7E): Sherburne W. Burnham, American astronomer (1838–1921)

Büsching (38S, 20E): Anton Friedrich Büsching, German geographer (1724–93)

Buys-Ballot (21N, 175E): Christoph H.D. Buys-Ballot, Dutch meteorologist (1817–90)

Byrd (83N, 0): Richard E. Byrd, American polar explorer (1888–1957)

Byrgius (25S, 65W): (Joost Bürgi) Swiss mathematician (1552–1632)

Cabannes (61S, 171W): Jean Cabannes, French physicist (1885–1959)

Caesar, Julius (9N, 15E): Roman statesman (100–44)

Cajori (48S, 168E): Florian Cajori, American mathematician (1859–1930)

Calippus (39N, 11E): Greek astronomer, fourth century BC

Campanus (28S, 28W): Johannes Campanus, Italian mathematician, thirteenth century

Campbell (45N, 152E) (1) Leon Campbell, American astronomer (1881–1951); (2) William W. Campbell, American astronomer (1862–1938)

Cannizzaro (55N, 100W): Stanislao Cannizzaro, Italian chemist (1826–1910)

Cannon (20N, 80E): Annie Jump Cannon, American astronomer (1863–1941)

Cantor (38N, 118E) (1) Georg Cantor, German mathematician (1845–1918); (2) Moritz B. Cantor, German mathematician (1829–1920)

Capella (8S, 36E): Martianus Capella, Roman encyclopaedist, fifth century AD

Capuanus (34S, 26W): (Francesco Capuano) Italian astronomer, fifteenth century

Cardanus (13N, 73W): (Geronimo Cardano) Italian mathematician (1501–76)

Carlini (34N, 24W): Francesco Carlini, Italian astronomer (1783–1862)

Carnot (52N, 144W): Nicolas L.S. Carnot, French physicist (1796–1832)

Carpatus, Montes (15N, 24W): 'Carpathian Mountains', named by Mädler (see below)

Carpenter (70N, 50W): James Carpenter, English astronomer (1826–75)

Carrington (44N, 62E): Richard C. Carrington, English astronomer (1826–75)

Carver (43S, 127E): George W. Carver, American chemist and botanist (?1864–1943)

Casatus (75S, 35W): (Paolo Casati) Italian mathematician (1617–1707)

Cassegrain (52S, 113E): Giovanni D. Cassegrain, French astronomer (1625–1712)

Cassini (40N, 5E) (1) Giovanni Domenico Cassini, Italian-born French astronomer (1625–1712);
  (2) Giacomo Cassini, Italian-born French astronomer (1677–1756), his son

Catalán (46S, 87W): Miguel A. Catalán, Spanish physicist (1894–1957)

Catharina (18S, 24E): (St Catherine) Christian virgin and martyr, fourth century AD

Caucasus, Montes (36N, 8E): 'Caucasus Mountains', named by Mädler (see below)

Cauchy (10N, 39E): Augustin Louis Cauchy, French mathematician (1789–1857)

Cavalerius (5N, 67W): (Francesco Cavalieri) Italian mathematician (1598–1647)

Cavendish (25S, 54W): Henry Cavendish, English chemist and physicist (1731–1810)

Cayley (4N, 15E): Arthur Cayley, English mathematician (1821–95)

Censorinus (0, 32E): Roman scholar, third century AD

Cepheus (41N, 46E): mythological King of Ethiopia (see constellation *Cepheus*)

Ceraski (49S, 141E): Vitold Karlovich Tserasky, Soviet astronomer (1849–1925)

Chacornac (30N, 32E): Jean Chacornac, French astronaut (1828–73)

Chaffee (39S, 155W): Roger B. Chaffee, American astronaut (1935–67)

Challis (78N, 9E): James Challis, English astronomer (1803–82)

Chamberlin (59S, 96E): Thomas C. Chamberlin, American geologist (1843–1928)

Champollion (37N, 175E): Jean-François Champollion, French Egyptologist (1790–1832)

Chandler (44N, 171E): Seth C. Chandler, American astronomer (1846–1913)

Chant (41S, 110W): Clarence A. Chant, Canadian astronomer (1865–1956)

Chaplygin (6S, 150E): Sergei A. Chaplygin, Soviet aerodynamics pioneer (1869–1942)

Chapman (50N, 101W): Sydney Chapman, English geophysicist (1888–1970)

Chappell (62N, 150W): James F. Chappell, American astronomer (1891–1964)

Charlier (36N, 132W): Carl W.L. Charlier, Swedish astronomer (1891–1964)

Chaucer (3N, 140W): Geoffrey Chaucer, English poet (?1340–1400)

Chauvenet (11S, 137E): William Chauvenet, American astronomer (1820–70)

Chebyshev (34S, 133W): Pafnuti L. Chebyshev, Russian mathematician (1821–94)

Chernyshev (47N, 174E): Nikolai G. Chernyshev, Soviet rocket engineer (1906–63)

Chevallier (45N, 52E): Temple Chevallier, French-born English astronomer (1794–1873)

Chladni (4N, 1E): Ernst Florens Friedrich Chladni, German physicist (1756–1827)

Chrétien (33S, 113E): Henri Chrétien, French optician (1879–1956)

Cichus (33S, 21W): (Cecco d'Ascoli) Italian astrologer (?1257–1327)

Clairaut (48S, 14E): Alexis Claude Clairaut, French mathematician (1713–65)

Clark (38S, 119E) (1) Alvan Clark, American lens-maker and astronomer (1804–87);

             (2) Alvan G. Clark, American lens-maker and astronomer (1832–97), his son

Clausius (37S, 44W): Rudolf Clausius, German physicist (1822–88)

Clavius (58S, 14W): Christopher Clavius, German astronomer (1537–1612)

Appendix I: Names of Craters on the Moon

Cleomedes (27N, 55E): Greek astronomer, first century AD (or later)
Coblentz (38S, 126E): William W. Coblentz, American astronomer (1873–1962)
Cockcroft (30N, 164W): John D. Cockroft, English physicist (1897–1967)
Colombo (15S, 46E): (Christopher Columbus) Genoese navigator (1451–1506)
Compton (55N, 104E) (1) Arthur H. Compton, American physicist (1892–1962);
(2) Karl T. Compton, American physicist (1887–1954), brothers
Comrie (23N, 113W): Leslie J. Comrie, New Zealand astronomer (1893–1950)
Comstock (21N, 122W): George C. Comstock, American astronomer (1855–1934)
Condorcet (12N, 70E): Jean de Caritat, Marquis de Condorcet, French philosopher and mathematician (1743–94)
Congreve (0, 168W): Sir William Congreve, English rocket pioneer (1772–1828)
Conon (22N, 2E): Greek astronomer and mathematician, third century BC
Cook (18S, 49E): James Cook, English mariner and explorer (1728–79)
Cooper (53N, 176E): John C. Cooper, American lawyer (1887–1967)
Copernicus (10N, 20W): (Mikołaj Kopernik) Polish astronomer (1473–1543)
Cordillera, Montes (27S, 85W): 'Cordillera Mountains'
Coriolis (0, 172E): Gaspard de Coriolis, French mathematician (1792–1843)
Crisium, Mare (18N, 58E)
Crocco (47S, 150E): Gaetano A. Crocco, Italian rocket engineer (1877–1968)
Crookes (11S, 165W): William Crookes, English physicist and chemist (1832–1919)
Crüger (17S, 67W): Peter Crüger, German mathematician (1580–1639)
Curie (23S, 92E): Pierre Curie, French chemist (1859–1906)
Curtius (67S, 5E): (Albert Curtz) German astronomer (1600–71)
Cusanus (72N, 70E): Nikolaus Krebs Cusanus, German mathematician (1401–64)

Cuvier (50S, 10E): Georges Cuvier, French naturalist (1769–1832)

Cyrano (20S, 157E): Cyrano de Bergerac, French poet and soldier (1619–55)

Cyrillus (13S, 24E): (St Cyril) Roman Catholic archbishop (376–444)

Cysatus (66S, 7W): (Jean-Baptiste Cysat) Swiss astronomer (1588–1657)

Daedalus (6S, 180): mythological inventor who devised wings for himself and his son Icarus

D'Alembert (52N, 164E): Jean le Rond d'Alembert, French mathematician (?1717–83)

Dalton (18N, 86W): John Dalton, English chemist and physicist (1766–1844)

Damoiseau (5S, 61W): Marie-Charles de Damoiseau, French astronomer (1768–1846)

Daniell (35N, 31E): John Frederic Daniell, English chemist (1790–1845)

Danjon (11S, 123E): André-Louis Danjon, French astronomer (1890–1967)

Dante (25N, 180): Italian poet (1265–1321)

Darney (15S, 24W): Maurice Darney, French astronomer (1882–1958)

D'Arrest (2N, 15E): Heinrich Ludwig d'Arrest, German astronomer (1822–75)

Darwin (20S, 69W): Charles Darwin, English naturalist (1809–82)

Das (14S, 152W): Amil K. Das, Indian astronomer (1902–61)

Da Vinci (9N, 45E): Leonardo da Vinci, Florentine painter (etc) (1452–1519)

Davisson (38S, 175W): Clinton J. Davisson, American physicist (1881–1958)

Davy (12S, 8W): Sir Humphry Davy, English chemist (1778–1829)

Dawes (17N, 26E): William R. Dawes, English astronomer (1799–1868)

Debes (29N, 52E): Ernest Debes, German cartographer (1840–1923)

Debye (50N, 177W): Peter J.W. Debye, Dutch physicist (1884–1966)

De Forest (76S, 162W): Lee De Forest, American radio engineer (1873–1961)

De Gasparis (26S, 50W): Annibale de Gasparis, Italian astronomer (1819–92)

Delambre (2S, 18E): Jean Baptiste Delambre, French astronomer (1749–1822)

De la Rue (67N, 55E): Warren de la Rue, English astronomer (1815–89)

Delaunay (22S, 3E): Charles Eugène Delaunay, French astronomer (1816–72)

Delisle (30N, 35W): Joseph Nicolas Delisle, French astronomer (1688–1768)

Dellinger (7S, 140E): John H. Dellinger, American radio engineer (1886–1962)

Delmotte (27N, 60E): Gabriel Delmotte, French astronomer (1876–1950)

Delporte (16S, 121E): Eugène J. Delporte, Belgian astronomer (1882–1955)

Deluc (55S, 3W): Jean André Deluc, Swiss geologist and meteorologist (1727–1817)

Dembowski (3N, 7E): Baron Ercole Dembowski, Italian astronomer (1815–81)

Democritus (62N, 35E): Greek philosopher, fifth or fourth century BC

Demonax (85S, 35E): Greek philosopher, second century BC

Denning (16S, 143E): William F. Denning, English astronomer (1848–1931)

De Roy (55S, 99W): Felix De Roy, Belgian amateur astronomer (1883–1942)

Descartes (12S, 16E): René Descartes, French scientist and philosopher (1596–1650)

Deseilligny (21N, 21E): Jules Deseilligny, French astronomer (1868–1918)

De Sitter (80N, 40E): Willem de Sitter, Dutch astronomer (1872–1934)

Deslandres (32S, 6W): Henri Deslandres, French astronomer (1853–1948)

Deutsch (24N, 111E): Armin J. Deutsch, American astronomer (1918–69)

De Vico (20S, 60W): Francesco de Vico, Italian astronomer (1805–48)

De Vries (20S, 177W): Hugo de Vries, Dutch botanist and geneticist (1848–1935)

Dewar (3S, 166E): Sir James Dewar, Scottish chemist and physicist (1842–1923)

Dionysius (3N, 17E): (St Dionysius) Athenian Christian convert, first century AD

Diophantus (28N, 34W): Greek mathematician, third century AD

Dirichlet (10N, 151W): Peter G.L. Dirichlet, German mathematician (1805–59)

Donati (21S, 5E): Giovanni Battista Donati, Italian astronomer (1826–73)

Donner (31S, 98E): Anders S. Donner, Finnish astronomer (1873–1949)

Doppelmayer (28S, 41W): Johann Doppelmayer, German astronomer (1671–1750)

Doppler (13S, 160W): Christian Johann Doppler, Austrian physicist (1803–53)

Dove (47S, 32E): Heinrich Wilhelm Dove, German physicist (1803–79)

Drebbel (41S, 49W): Cornelis Drebbel, Dutch inventor (1572–1634)

Dreyer (10N, 97E): John L.E. Dreyer, Danish astronomer (1852–1926)

Drude (39S, 91W): Paul Karl Ludwig Drude, German physicist (1863–1906)

Dryden (33S, 157W): Hugh L. Dryden, American physicist (1898–1965)

Drygalski (80S, 55W): Erich von Drygalski, German geographer (1865–1949)

Dubiago (4N, 70E): Dmitri I. Dubiago, Russian astronomer (1850–1918)

Dufay (5N, 170E): Jean C.B. Dufay, French astronomer (1896–1967)

Dunér (45N, 179E): Nils Christofer Dunér, Swedish astronomer (1839–1914)

Dyson (61N, 121W): Sir Frank W. Dyson, English astronomer (1868–1939)

Dziewulski (21N, 99E): Wladyslaw Dziewulski, Polish astronomer (1878–1962)

Eddington (22N, 72W): Sir Arthur S. Eddington, English astronomer (1882–1944)

Egede (49N, 11E): Hans Egede, Norwegian missionary to Eskimos (1686–1758)

Ehrlich (41N, 172W): Paul Ehrlich, German bacteriologist (1854–1915)

Eichstädt (23S, 80W): Lorenz Eichstädt, German mathematician (1596–1660)

Eijkman (62S, 141W): Christiaan Eijkman, Dutch physician (1858–1930)

Eimmart (24N, 65E): Georg Christoph Eimmart, German astronomer (1638–1705)

Einstein (18N, 86W): Albert Einstein, German physicist (1879–1955)

Einthoven (5S, 110E): Willem Einthoven, Dutch physiologist (1860–1927)

Elger (35S, 30W): Thomas G. Elger, English astronomer (1838–97)

Ellerman (26S, 121W): Ferdinand Ellerman, American astronomer (1869–1940)

Ellison (55N, 108W): Mervyn A. Ellison, Irish astronomer (1909–63)

Elvey (9N, 101W): Christian T. Elvey, American astronomer (1899–1970)

Emden (63N, 176W): Robert Emden, German astrophysicist (1862–1940)

Emley (38S, 28W): Edward F. Emley, English astronomer (1920–80)

Encke (5N, 37W): Johann Encke, German astronomer (1791–1865) (see *Encke's Comet*)

Endymion (55N, 55E): mythological youth visited by moon goddess Selene nightly

Engelhardt (5N, 159W): Vasily P. Engelhardt, Russian astronomer (1828–1915)

Eötvös (36S, 134E): Baron Roland Eötvös, Hungarian physicist (1848–1919)

Epidemiarum, Palus (31S, 26W): 'Marsh of Epidemics', named by Schmidt (see below)

Epigenes (73N, 4W): Greek astronomer, third century BC

Epimenides (41S, 30W): Cretan philosopher and poet, seventh century BC

Eratosthenes (15N, 11W): Greek astronomer and geographer, third century BC

Erro (6N, 93E): Luis E. Erro, Mexican astronomer and novelist (1897–1955)

Esnault-Pelterie (47N, 142W): Robert Esnault-Pelterie, French aviation pioneer (1881–1957)

Espin (28N, 109E): Thomas H.E.C. Espin, English amateur
astronomer (1858–1934)

Euclides (7S, 29W): (Euclid) Greek geometer, third century BC

Euctemon (80N, 40E): Greek astronomer, fifth century BC

Eudoxus (44N, 16E): Greek astronomer, fourth century BC

Euler (23N, 29W): Leonhard Euler, Swiss mathematician
(1707–83)

Evans (10S, 134W): Sir Arthur Evans, English archaeologist
(1851–1941)

Evershed (36N, 160W): John Evershed, English astronomer
(1864–1956)

Fabricius (43S, 42E): David Fabricius, German astronomer
(1564–1617)

Fabry (43N, 100E): Charles Fabry, French physicist (1867–1945)

Faraday (42S, 8E): Michael Faraday, English chemist and
physicist (1791–1867)

Fauth (6N, 20W): Philipp Johann Heinrich Fauth, German
astronomer (1867–1943)

Faye (21S, 4E): Hervé Faye, French astronomer (1814–1902)

Fechner (59S, 125E): Gustave Theodor Fechner, German
physicist (1801–87)

Fecunditatis, Mare (4S, 51E)

Fenyi (45S, 105W): (Father) J. Fenyi, Hungarian astronomer
(1845–1927)

Fermat (23S, 20E): Pierre de Fermat, French mathematician
(1601–65)

Fermi (20S, 123E): Enrico Fermi, Italian physicist (1901–54)

Fernelius (38S, 5E): (Jean Fernel) French astronomer (1497–1558)

Fersman (18N, 126W): Alexander Y. Fersman, Soviet geochemist
(1883–1945)

Feuillée (27N, 10W): Louis Feuillée, French astronomer
(1660–1732)

Firmicus (7N, 64E): Firmicus Maternus, Latin writer, fourth
century AD

Firsov (4N, 112E): Georgy F. Firsov, Soviet rocket engineer
(1917–60)

Fitzgerald (27N, 172W): George Francis Fitzgerald, Irish physicist
(1851–1901)

Fizeau (58S, 133W): Armand H.L. Fizeau, French physicist
(1819–96)

Flammarion (3S, 4W): Camille Flammarion, French astronomer (1842–1925)

Flamsteed (5S, 44W): John Flamsteed, English astronomer (1646–1719)

Fleming (15N, 109E) (1) Sir Alexander Fleming, Scottish bateriologist (1881–1955);

                    (2) Williamina P. Fleming, American astronomer (1857–1911)

Focas (34S, 94W): Ionnas Focas, Greek/French astronomer (1908–69)

Fontana (16S, 57W): Francesco Fontana, Italian astronomer (?1585–1656)

Fontenelle (63N, 19W): Bernard le Bovier de Fontenelle, French savant (1657–1757)

Foucault (50N, 40W): Léon Foucault, French physicist (1819–68)

Fourier (30S, 53W): Jean Baptiste Fourier, French physicist (1768–1830)

Fowler (43N, 145W) (1) Alfred Fowler, English astrophysicist (1868–1940);

                (2) Sir Ralph H. Fowler, English mathematician (1889–1944)

Fracastorius (21S, 33E): (Girolamo Fracastoro) Italian astronomer (1483–1553)

Fra Mauro (6S, 17W): Venetian geographer (died 1459)

Franklin (39N, 48E): Benjamin Franklin, American statesman and scientist (1706–90)

Franz (16N, 40E): Julius H. Franz, German astronomer (1847–1913)

Fraunhofer (39S, 59E): Joseph von Fraunhofer, German optician (1787–1826)

Freundlich (25N, 171E): Erwin (Finlay-)Freundlich, German-British astronomer (1885–1964)

Fridman (13S, 127W): Alexander A. Fridman, Soviet mathematician (1888–1925)

Frigoris, Mare (55N, 0)

Froelich (80N, 110W): Jack E. Froelich, American rocket engineer (1921–67)

Frost (37N, 119W): Edwin B. Frost, American astronomer (1866–1935)

Furnerius (36S, 60E): (Georges Furner) French mathematician (died ?1643)

Gadomski (36N, 147W): Jan Gadomski, Polish astronomer (1889–1966)

Gagarin (20S, 150E): Yury A. Gagarin, Soviet cosmonaut (1934–68)

Galilaei (10N, 63W): (Galileo Galilei) Italian astronomer (1564–1642)

Galle (56N, 22E): Johann Gottfried Galle, German astronomer (1812–1910)

Galois (16S, 153W): Evariste Galois, French mathematician (1811–32)

Galvani (49N, 85W): Luigi Galvani, Italian physicist (1737–98)

Gambart (1N, 15W): Jean Félix Gambart, French astronomer (1800–36)

Gamow (65N, 143E): George Gamow (Georgy Antonovich Gamov), Russian-born American physicist (1904–68)

Ganswindt (79S, 110E): Herman Ganswindt, German rocket pioneer (1856–1934)

Garavito (48S, 157E): J. Garavito, Colombian astronomer (1865–1920)

Gärtner (60N, 34E): Christian Gärtner, German geologist (1750–1813)

Gassendi (18S, 40W): Pierre Gassendi, French philosopher and savant (1592–1655)

Gaudibert (11S, 37E): Casimir Gaudibert, French astronomer (1823–1901)

Gauricus (34S, 12W): (Luca Gaurico) Italian astronomer (1476–1558)

Gauss (36N, 80E): Karl Gauss, German astronomer (1777–1855) (see *Gaussia*)

Gavrilov (17N, 131E): Alexander I. Gavrilov, Soviet rocket engineer (1884–1955)

Gay-Lussac (14N, 21W): Joseph Louis Gay-Lussac, French chemist (1778–1850)

Geber (20S, 14E): (Jabir ibn-Aflah) Arab astronomer, twelfth century

Geiger (14S, 158E): Hans Geiger, German physicist (1882–1945)

Geminus (35N, 57E): Greek astronomer, first century BC

Gemma Frisius (34S, 14E): Reiner Gemma, Dutch geographer (1508–55)

Gerasimovič (23S, 124W): Boris P. Gerasimovich, Soviet astronomer (1889–1937)

Gernsback (36S, 99E): Hugo Gernsback, American science fiction writer (1884–1967)

Gibbs (19S, 83E): Josiah Willard Gibbs, American physicist (1839–1903)

Gilbert (4S, 75E): Grove Karl Gilbert, American geologist (1843–1918)

Gill (64S, 75E): Sir David Gill, Scottish astronomer (1843–1914)

Ginzel (14N, 97E): Friedrich K. Ginzel, Austrian astronomer (1850–1926)

Gioja (89N, 9W): Flavio Gioja, Italian mariner, fourteenth century

Glasenap (2S, 138E): Sergei P. Glazenap, Soviet astronomer (1848–1937)

Goclenius (10S, 45E): (Rudolf Goeckel) German philosopher (1547–1628)

Goddard (16N, 90E): Robert Hutchings Goddard, American physicist (1882–1945)

Godin (2N, 10E): Louis Godin, French astronomer (1704–60)

Goldschmidt (75N, 0): Hermann Goldschmidt, German amateur astronomer (1802–66)

Golitzyn (25S, 105W): Boris B. Golitsyn, Russian physicist (1862–1916)

Golovin (40N, 161E): Nicholas E. Golovin, American rocket engineer (1912–69)

Goodacre (33S, 14E): Walter Goodacre, English astronomer (1856–1938)

Gould (19S, 17W): Benjamin A. Gould, American astronomer (1824–96)

Grachev (3S, 108W): Andrei D. Grachev, Soviet rocket design engineer (1900–64)

Graff (43S, 88W): Kasimir Romuald Graff, German astronomer (1878–1950)

Green (4N, 133E): George Green, English mathematician (1793–1841)

Gregory (2N, 127E): James Gregory, Scottish mathematician (1638–75)

Grimaldi (6S, 68W): Francesco Maria Grimaldi, Italian physicist (?1618–63)

Grissom (45S, 160W): Virgil I. Grissom, American astronaut (1926–67)

Grotrian (66S, 128E): W. Grotrian, German astronomer (1890–1954)

Grove (40N, 33E): Sir William Robert Grove, English physicist (1811–96)

Gruemberger (68S, 10W): Christoph Gruemberger, Austrian astronomer (1561–1636)

Gruithuisen (33N, 40W): Franz von Gruithuisen, German physician (1774–1852)

Guericke (12S, 14W): Otto von Guericke, German physicist (1602–86)

Gullstrand (45N, 130W): Allvar Gullstrand, Swedish ophthalmologist (1862–1930)

Gum (40S, 89E): Colin S. Gum, Australian astronomer (1924–60) (see *Gum Nebula*)

Gutenberg (8S, 41E): Johann Gutenberg, German inventor of printing (?1400–?1468)

Guthnick (48S, 94W): Paul Guthnick, German astronomer (1879–1947)

Guyot (11N, 117E): Arnold H. Guyot, Swiss-born American geographer (1807–84)

Haemus, Montes (16N, 14E): 'Haemus Mountains', old name for Balkan Mountains

Hagecius (60S, 46E): (Thaddaeus Hayek) Czech mathematician (1525–1600)

Hagen (56S, 135E): Johannes Georg Hagen, Austrian astronomer (1847–1930)

Hahn (31N, 74E): Friedrich, Graf von Hahn, German astronomer (1741–1805)

Haidinger (39S, 25W): Wilhelm Karl von Haidinger, Austrian geologist (1795–1871)

Hainzel (41S, 34W): Paul Hainzel, German astronomer (died ?1570)

Hale (74S, 90E) (1) George E. Hale, American astronomer (1868–1938);
(2) William Hale, English scientist (1797–1870)

Hall (34N, 37E): Asaph Hall, American astronomer (1829–1907) (see *Deimos*)

Halley (8S, 6E): Edmond Halley, English astronomer (1656–1742) (see *Halley's Comet*)

Hamilton (44S, 83E): Sir William R. Hamilton, Irish mathematician (1805–65)

Hanno (57S, 73E): Carthaginian navigator, sixth or fifth century BC

Hansen (44S, 83E): Peter Andreas Hansen, Danish astronomer (1795–1874)

Hansky (10S, 97E): Alexei P. Hansky (Gansky), Russian astronomer (1870–1908)

Hansteen (12S, 52W): Christopher Hansteen, Norwegian astronomer (1784–1873)

Harding (43N, 70W): Karl Ludwig Harding, German astronomer (1765–1834) (see *Juno*)

Harpalus (53N, 43W): Greek astronomer, fourth century BC

Harriot (33N, 114E): Thomas Harriot, English mathematician (1560–1621)

Hartmann (3N, 135E): Johannes F. Hartmann, German astronomer (1865–1936)

Hartwig (7S, 82W): Carl E. Hartwig, German astronomer (1851–1923)

Harvey (19N, 147W): William Harvey, English physician (1578–1657)

Hase (30S, 63E): Johann M. Hase, German mathematician (1684–1742)

Hauet (37S, 18W): Pierre Hauet, French astronomer (1878–1933)

Hausen (65S, 90W): Christian A. Hausen, German astronomer (1693–1743)

Hayford (13N, 176W): John F. Hayford, American geodesist (1868–1925)

Hayn (63N, 85E): Friedrich Hayn, German astronomer (1863–1928)

Healy (32N, 111W): Roy Healy, American rocket designer (1915–68)

Heaviside (10S, 167E): Oliver Heaviside, English physicist (1850–1925)

Hecataeus (23S, 84E): Greek traveller and historian, fifth century BC

Heinsius (39S, 18W): Gottfried Heinsius, German astronomer (1709–69)

Heis (32N, 32W): Eduard Heis, German astronomer (1806–77)

Helberg (22N, 102W): Robert J. Helberg, American aeronautical engineer (1906–67)

Helicon (40N, 23W): Greek mathematician and astronomer, fourth century BC

Hell (32S, 8W): Maximilian Hell, Hungarian astronomer (1720–92)

Helmholtz (72N, 78E): Hermann von Helmholtz, German physicist (1821–94)

Henderson (5N, 152E): Thomas Henderson, Scottish astronomer (1798–1844)

Hendrix (48S, 161W): Don O. Hendrix, American optician (1905–61)

Henry, Paul (24S, 57W): French astronomer (1848–1905)

Henry, Prosper (24S, 59W): French astronomer (1849–1903), brother of Paul

Henyey (13N, 152W): Louis G. Henyey, American astronomer (1910–70)

Heraclitus (49S, 6E): Greek philosopher, sixth to fifth century BC

Hercules (46N, 39E): mythological hero and 'superman' (see constellation *Hercules*)

Herigonius (13S, 34W): (Pierre Hérigone) French mathematician (died ?1644)

Hermann (1S, 57W): Jacob Hermann, Swiss mathematician (1678–1733)

Hermite (85N, 90W): Charles Hermite, French mathematician (1822–1901)

Herodotus (23N, 50W): Greek historian, fifth century BC

Herschel (6S, 2W): Sir William Herschel, German-born English astronomer (1738–1822) (see *Uranus*)

Herschel, C. (34N, 31W): Caroline L. Herschel, English astronomer (1750–1848), sister of William (see also *Herschel-Rigollet Comet*)

Herschel, J. (62N, 41W): John Herschel, English astronomer (1792–1871), son of William

Hertz (14N, 104E): Heinrich Rudolf Hertz, German physicist (1857–94)

Hertzsprung (0, 130W): Ejnar Hertzsprung, Danish astronomer (1873–1967)

Hesiodus (29S, 16W): (Hesiod) Greek poet, eighth century BC

Hess (54S, 174E) (1) Harry H. Hess, American geologist (1906–69);
(2) Victor Franz Hess, Austrian physicist (1883–1964)

Hevelius (2N, 67W): Johannes Hevelius, German astronomer (1611–87)

Heymans (75N, 145W): Corneille Heymans, Belgian physiologist (1892–1968)

Hilbert (18S, 108E): David Hilbert, German mathematician
(1862–1943)

Hind (8S, 7E): John Russell Hind, English astronomer (1823–95)

Hippalus (25S, 30W): Greek navigator, second century BC

Hipparchus (6S, 5E): Greek astronomer, second century BC

Hippocrates (71N, 146W): Greek physician (?460–?377)

Hirayama (6S, 93E) (1) Hirayama Seiji, Japanese astronomer
(1874–1943);
(2) Shin Hirayama, Japanese astronomer
(1867–1945)

Hoffmeister (15N, 137E): Kuno Hoffmeister, German astronomer
(1892–1968)

Hogg (34N, 122E) (1) Arthur R. Hogg, Australian astronomer
(1903–66);
(2) Frank S. Hogg, Canadian astronomer
(1904–51)

Hohmann (18S, 94W): Walter Hohmann, German engineer
(1880–1945)

Holden (19S, 63E): Edward S. Holden, American astronomer
(1846–1914)

Holetschek (28S, 151E): Johann Holetschek, Austrian astronomer
(1846–1923)

Hommel (54S, 33E): Johann Hommel, German mathematician
and astronomer (1518–62)

Hooke (41N, 55E): Robert Hooke, English experimental
philosopher (1635–1703)

Horrebow (59N, 41W): Peder Horrebow, Danish mathematician
(1679–41)

Horrocks (4S, 6E): Jeremiah Horrocks, English astronomer
(?1617–41)

Hortensius (6N, 28W): Martin van den Hove Hortensius, Dutch
astronomer (1605–39)

Houzeau (18S, 124W): Jean Charles Houseau de Lehaye, Belgian
astronomer (1820–88)

Hubble (22N, 87E): Edwin P. Hubble, American astronomer
(1889–1953) (see *Hubble's Variable Nebula*)

Huggins (41S, 2W): Sir William Huggins, English astronomer
(1824–1910)

Humboldt, W. (27S, 81E): Wilhelm von Humboldt, German
statesman and philologist (1767–1835), brother of Alexander
von Humboldt (see *Humboldtianum, Mare*)

Humboldtianum, Mare (55N, 75E)

Humorum, Mare (23S, 38W)

Hutton (37N, 169E): James Hutton, Scottish geologist (1726–97)

Hyginus (8N, 6E): Gaius Julius Hyginus, Latin author, first century BC

Hyginus, Rima (8N, 6E): 'Hyginus' Rill', runs by crater Hyginus (see above)

Hypatia (4S, 23E): Greek philosopher (murdered 415 AD), daughter of Theon (B) (see below)

Ibn Yunus (14N, 91E): Ali ibn-Abd Rachman, Arab astronomer (950–1009)

Icarus (6S, 173W): mythological youth who flew near the Sun (see asteroid *Icarus*)

Ideler (49S, 22E): Christian Ludwig Ideler, German astronomer (1766–1846)

Idelson (81S, 114E): Naum I. Idelson, Soviet astronomer (1885–1951)

Imbrium, Mare (36N, 16W)

Ingalls (4S, 153W): Albert L. Ingalls, American amateur telescope-maker (1888–1958)

Ingenii, Mare (35S, 164E)

Inghirami (48S, 70W): Giovanni Inghirami, Italian astronomer (1779–1851)

Innes (28N, 119E): Robert T.A. Innes, Scottish astronomer (1861–1933)

Iridum, Sinus (45N, 32W)

Isidorus (8S, 33E): (St Isidore) Spanish prelate and scholar (?560–636)

Izsak (23S, 117E): Imre Iszak, Hungarian-born American astronomer (1929–65)

Jackson (22N, 163W): John Jackson, Scottish astronomer (1887–1958)

Jacobi (57S, 12E): Karl Gustav Jakob Jacobi, German mathematician (1804–51)

Jansen (14N, 29E): Zacharias Janssen, Dutch optician (died ?1690)

Jansky (8N, 87E): Karl Jansky, American radiophysicist (1905–50)

Janssen (46S, 40E): Pierre Jules César Janssen, French astronomer (1824–1907)

Jeans (53S, 91W): Sir James Jeans, English physicist and astronomer (1877–1946)

Jenner (42S, 96E): Edward Jenner, English physicist (1749–1823)

Joffe (15S, 129W): Abram F. Joffe, Soviet physicist (1880–1960)

Joliot (27N, 93E): Frédéric Joliot-Curie, French physicist (1900–58)

Joule (27N, 144W): James P. Joule, English physicist (1818–89)

Jura, Montes (46N, 38W): 'Jura Mountains', named by Debes (see above)

Kaiser (36S, 6E): Frederick Kaiser, Dutch astronomer (1808–72)

Kamerlingh Onnes (15N, 116W): Heike Kamerlingh Onnes, Dutch physicist (1853–1926)

Kane (63N, 26E): Elisha K. Kane, American explorer (1820–57)

Kant (11S, 20E): Immanuel Kant, German philosopher (1724–1804)

Kapteyn (13S, 70E): Jacobus Kapteyn, Dutch astronomer (1851–1922) (see *Kapteyn's Star*)

Karpinsky (73N, 166E): Alexei P. Karpinsky, Soviet geologist (1846–1936)

Kästner (6S, 80E): Abraham Gotthelf Kästner, German mathematician (1719–1800)

Kearons (12S, 113W): William M. Kearons, American astronomer (1878–1948)

Keeler (10S, 162E): James E. Keeler, American astronomer (1857–1900)

Kekulé (16N, 138W): Friedrich August Kekule, German chemist (1829–96)

Kepler (8N, 38W): Johannes Kepler, German astronomer (1571–1630) (see *Kepler's Star*)

Kibalchich (2N, 147W): Nikolai I. Kibalchich, Russian rocket engineer (1853–81)

Kidinnu (36N, 123E): Babylonian astronomer and mathematician, fourth century BC

Kies (26S, 23W): Johann Kies, German mathematician (1713–81)

Kimura (57S, 118E): Hisachi Kimura, Japanese astronomer (1870–1943)

Kinau (60S, 15E): C.A. Kinau, German botanist (died ?1850)

King (5N, 120E)   (1) Arthur S. King, American physicist (1876–1957);
                  (2) Edward S. King, American astronomer (1861–1931)

Kirch (39N, 6W): Gottfried Kirch, German astronomer (1639–1710)

Kirchhoff (30N, 39E): Gustav Robert Kirchhoff, German physicist (1824–87)

Kirkwood (69N, 157W): Daniel Kirkwood, American astronomer (1814–95)

Klaproth (68S, 22W): Martin Heinrich Klaproth, German chemist (1743–1817)

Kleimenov (33S, 141W): Ivan T. Kleimenov, Soviet rocket engineer (1898–1938)

Klein (12S, 3E): Hermann J. Klein, German astronomer (1844–1914)

Klute (37N, 142W): Daniel O. Klute, American rocket engineer (1921–64)

Koch (43S, 150E): Robert Koch, German physician (1843–1910)

Kohlschütter (15N, 154E): Arnold Kohlschütter, German astronomer (1883–1969)

Kolhörster (10N, 115W): Werner Kolhörster, German physicist (1887–1946)

Komarov (25N, 153E): Vladimir M. Komarov, Soviet cosmonaut (1927–67)

Kondratyuk (15S, 115E): Yuri V. Kondratyuk, Soviet rocket pioneer (1897–1942)

König (24S, 25W): Rudolf König, Austrian astronomer (1865–1927)

Konstantinov (20N, 159E): Konstantin I. Konstantinov, Russian rocket engineer (1817–71)

Kopff (17S, 90W): August Kopff, German astronomer (1882–1960)

Korolev (5S, 157W): Sergei P. Korolev, Soviet rocket designer (1906–66)

Kostinsky (14N, 118E): Sergei K. Kostinsky, Soviet astronomer (1867–1936)

Kovalevskaya (31N, 129W): Sofya V. Kovalevskaya, Russian mathematician (1850–91)

Kovalsky (22S, 101E): Marian A. Kovalsky, Russian astronomer (1821–84)

Krafft (13N, 73W): Wolfgang Ludwig Krafft, German astronomer (1743–1814)

Krasnov (31S, 89W): Alexander V. Krasnov, Russian astronomer (1866–1907)

Krasovsky (4N, 176W): Feodosi N. Krasovsky, Soviet astronomer (1878–1948)

Krieger (29N, 46W): Johann Krieger, German astronomer (1865–1902)

Krylov (9S, 157W): Alexei N. Krylov, Soviet mathematician
(1863–1945)

Kugler (53S, 104E): F.X. Kugler, German linguist and historian
(1862–1929)

Kulik (42N, 155W): Leonid A. Kulik, Soviet mineralogist
(1883–1942)

Kunowsky (3N, 32W): Georg K.F. Kunowsky, German
astronomer (1786–1846)

Kuo Shou Ching (8N, 134W): Chinese astronomer, fourteenth
century

Kurchatov (39N, 142E): Igor V. Kurchatov, Soviet nuclear
physicist (1902–60)

Lacaille (24S, 1E): Nicolas Louis de Lacaille, French astronomer
(1713–62)

Lacchini (41N, 107W): Giovanni Lacchini, Italian amateur
astronomer (1884–1967)

La Condamine (53N, 28W): Charles de la Condamine, French
geographer (1701–74)

Lacroix (38S, 59W): Sylvestre François Lacroix, French
mathematician (1765–1843)

Lade (1S, 10E): Heinrich von Lade, German astronomer
(1817–1904)

La Hire, Mons (28N, 25W): 'La Hire Mountain', named for
Philippe de la Hire, French astronomer (1640–1718)

Lalande (4S, 8W): Joseph Lalande, French astronomer
(1732–1807) (see star *Lalande* 21885)

Lamarck (23S, 69W): Jean Baptiste Lamarck, French naturalist
(1744–1829)

Lamb (43S, 101E): Sir Horace Lamb, English mathematician
(1849–1934)

Lambert (26N, 21W): Johann Heinrich Lambert, German
physicist (1728–77)

Lamé (15S, 65E): Gabriel Lamé, French mathematician
(1795–1870)

Lamèch (43N, 13E): Felix C. Lamèch, French astronomer
(1894–1962)

Lamont (5N, 23E): Johann von Lamont, Scots-German
astronomer (1805–79)

Lampland (31S, 131E): Carl O. Lampland, American astronomer
(1873–1951)

Landau (42N, 119W): Lev D. Landau, Soviet physicist (1908–68)

Lane (9S, 132E): Jonathan H. Lane, American physicist (1819–80)

Langemak (10S, 119E): Georgy E. Langemak, Soviet rocket designer (1898–1938)

Langevin (44N, 162E): Paul Langevin, French physicist (1872–1946)

Langley (52N, 87W): Samuel P. Langley, American astronomer (1834–1906)

Langmuir (36S, 129W): Irving Langmuir, American chemist (1881–1957)

Langrenus (9S, 61E): (Michel van Langren) Belgian mathematician (?1600–?1675)

Lansberg (0, 26W): Philippe van Lansberg, Belgian physician (1561–1632)

La Pérouse (10S, 78E): Jean François de Galaup, Comte de la Pérouse, French explorer (1741–88)

Larmor (32N, 180): Sir Joseph Larmor, Irish mathematician (1857–1942)

Lassell (16S, 8W): William Lassell, English astronomer (1799–1880)

Laue (29N, 96W): Max von Laue, German physicist (1879–1960)

Lauritsen (27S, 96E): Charles Christian Lauritsen, Danish physicist (1892–1968)

Lavoisier (36N, 70W): Antoine Laurent Lavoisier, French chemist (1743–94)

Leavitt (46S, 140W): Henrietta S. Leavitt, American astronomer (1868–1921)

Lebedev (48S, 108E): Pyotr N. Lebedev, Russian physicist (1866–1911)

Lebedinsky (8N, 165W): Alexander I. Lebedinsky, Soviet astrophysicist (1913–67)

Lee (31S, 41W): John Lee, English astronomer (1783–1866)

Leeuwenhoek (30S, 179W): Anton van Leeuwenhoek, Dutch naturalist (1632–1723)

Legendre (29S, 70E): Adrien Marie Legendre, French mathematician (1752–?1833)

Legentil (73S, 80E): Guillaume Legentil, French astronomer (1725–92)

Lehmann (40S, 56W): Jacob Lehmann, German astronomer (1800–63)

Leibnitz (38S, 178E): Baron Gottfried von Leibnitz, German philosopher (1646–1716)

Lemaître (62S, 150W): (Abbé) George Lemaître, Belgian astrophysicist (1894–1966)

Lemonnier (26N, 31E): Pierre Charles Lemonnier, French astronomer (1715–99)

Leonov (19N, 148E): Alexei A. Leonov, Soviet cosmonaut (born 1934)

Lepaute (33S, 34W): Nicole Reine Lepaute, French arithmetician (1723–88)

Letronne (10S, 43W): Jean Antoine Letronne, French archaeologist (1787–1848)

Leucippus (29N, 116W): Greek philosopher, fifth century BC

Leuschner (1N, 109W): Armin O. Leuschner, American astronomer (1868–1953)

Leverrier (40N, 20W): Urbain Leverrier, French astronomer (1811–77) (see *Neptune*)

Levi-Civita (24S, 143E): Tullio Levi-Civita, Italian mathematician (1873–1942)

Lewis (19S, 114W): Gilbert N. Lewis, American chemist (1875–1946)

Lexell (36S, 4W): Anders J. Lexell, Finnish astronomer (1740–84)

Ley (43N, 154E): Willy Ley, German rocket pioneer (1906–69)

Liapunov (27N, 88E): Alexander M. Liapunov, Russian mathematician (1857–1918)

Licetus (47S, 6E): (Fortunio Liceti) Italian philosopher (1577–1657)

Lichtenberg (32N, 68W): Georg Christoph Lichtenberg, German physicist (1742–99)

Lick (12N, 53E): James Lick, American financier and philanthropist (1796–1876)

Liebig (24S, 48W): Baron Justus von Liebig, German chemist (1803–73)

Lilius (54S, 6E): (Luigi Giglio) Italian philosopher (died 1576)

Lindenau (32S, 25E): Baron Bernhard von Lindenau, German astronomer (1779–1854)

Linné (28N, 12E): (Linnaeus) Swedish botanist (1707–78)

Lippershey (26S, 10W): Hans Lippershey, Dutch spectacle-maker (died ?1619)

Littrow (22N, 31E): Joseph von Littrow, Austrian astronomer (1781–1840)

Lobachevsky (9N, 112E): Nikolai I. Lobachevsky, Russian mathematician (1793–1856)

Lockyer (46S, 37E): Sir Joseph Norman Lockyer, English
astronomer (1836–1920)
Lodygin (18S, 147W): Alexander N. Lodygin, Russian electrical
engineer (1847–1923)
Lohrmann (1S, 67W): Wilhelm G. Lohrmann, German
astronomer (1796–1840)
Lohse (14S, 60E): Oswald Lohse, German astronomer
(1845–1915)
Lomonosov (28N, 98E): Mikhail V. Lomonosov, Russian scientist
(1711–65)
Longomontanus (50S, 21W): Christian Severin Longomontanus,
Danish astronomer (1562–1647)
Lorentz (34N, 100W): Hendrik Antoon Lorentz, Dutch physicist
(1853–1928)
Love (6S, 129E): Augustus E.H. Love, English geophysicist
(1863–1940)
Lovelace (82N, 107W): William R. Lovelace II, American
physician (1907–65)
Lovell (39S, 149W)   (1) Sir Bernard Lovell, English astronomer
                          (born 1913);
                      (2) James A. Lovell, American astronaut
                          (born 1928)
Lowell (13S, 103W): Percival Lowell, American astronomer
(1855–1916) (see *Pluto*)
Lubbock (4S, 42E): Sir John William Lubbock, English
astronomer (1803–65)
Lubiniezky (18S, 24W): Stanislau Lubiniezky, Polish astronomer
(1623–75)
Lucretius (9S, 121W): Roman philosophical poet, first century BC
Lundmark (39S, 152E): Knut E. Lundmark, Swedish astronomer
(1889–1958)
Luther (33N, 24E): Robert Luther, German astronomer
(1822–1900)
Lütke (17S, 123E): Fyodor P. Lütke, Russian navigator
(1797–1882)
Lyell (14N, 41E): Sir Charles Lyell, Scottish geologist (1797–1875)
Lyman (65S, 162E): Theodore Lyman, American physicist
(1874–1954)
Lyot (48S, 88W): Bernard Ferdinand Lyot, French astronomer
(1897–1952)

Mach (18N, 149W): Ernst Mach, Austrian physicist (1838–1916)

McKellar (16S, 171W): Andrew McKellar, Canadian astronomer (1910–60)

McLaughlin (47N, 93W): Dean B. McLaughlin, American astronomer (1901–65)

Maclaurin (2S, 68E): Colin Maclaurin, Scottish mathematician (1698–1746)

Maclear (11N, 20E): Sir Thomas Maclear, English-South African astronomer (1794–1879)

McMath (15N, 167W) (1) Francis C. McMath, American engineer and astronomer (1867–1938);

  (2) Robert R. McMath, American astronomer (1891–1962)

McNally (22N, 127W): Paul A. McNally, American astronomer (1890–1955)

Macrobius (21N, 46E): Ambrosius Macrobius, Latin grammarian, fourth to fifth century AD

Mädler (11S, 30E): Johann Heinrich von Mädler, German astronomer (1794–1874)

Magelhaens (12S, 44E): (Magellan) Portuguese navigator (?1480–1521)

Maginus (50S, 6W): (Giovanni Magini) Italian astronomer (1555–1617)

Main (82N, 3E): Robert Main, English astronomer (1808–78)

Mairan (42N, 43W): Jean de Mairan, French astronomer (1678–1771)

Maksutov (41S, 169W): Dmitri D. Maksutov, Soviet optician (1896–1964)

Malyi (22N, 105E): Alexander L. Malyi, Soviet rocket engineer (1907–61)

Mandelstam (4N, 156E): Leonid I. Mandelstam, Soviet physicist (1879–1944)

Manilius (15N, 9E): Roman poet, first century BC or AD

Manners (5N, 20E): Russell Henry Manners, English astronomer (1800–70)

Manzinus (68S, 25E): (Carlo Manzini) Italian astronomer (1599–1677)

Maraldi (19N, 35E): Giovanni Domenico Maraldi, Italian astronomer (1709–88)

Marconi (9S, 145E): Guglielmo Marconi, Italian electrical engineer (1874–1937)

Marginis, Mare (13N, 87E)

Marinus (50S, 75E): Greek geographer, second century AD
Mariotte (29S, 140W): Edmé Mariotte, French physicist
(?1620–84)
Marius (12N, 51W): (Simon Mayr) German astronomer
(1570–1624)
Markov (54N, 62W)   (1) Andrei A. Markov, Russian
mathematician (1856–1922);
(2) Alexander V. Markov, Soviet
astrophysicist (1897–1968)
Maskelyne (2N, 30E): Nevil Maskelyne, English astronomer
(1732–1811)
Mason (43N, 30E): Charles Mason, English astronomer (1730–87)
Maunder (14S, 94W)   (1) Annie S.D.R. Maunder, Irish
astronomer (1858–1947);
(2) Edward W. Maunder, Irish
astronomer (1851–1928), husband
and wife
Maurolycus (42S, 14E): (Francesco Maurolico) Italian
mathematician (1494–1575)
Maury (37N, 40E)   (1) Matthew F. Maury, American
oceanographer (1806–73);
(2) Antonia C. Maury, American astronomer
(1866–1952)
Maxwell (30N, 99E): James Clerk Maxwell, Scottish physicist
(1831–79)
Mayer, C. (63N, 17E): Christian Mayer, German astronomer
(1719–83)
Mayer, T. (16N, 29W): Tobias Mayer, German astronomer
(1723–62)
Medii, Sinus (0, 0)
Mee (44S, 35W): Arthur B.P. Mee, Scottish astronomer
(1860–1926)
Mees (14N, 96W): C.E. Kenneth Mees, British-American
photographer (1882–1960)
Meggers (24N, 123E): William F. Meggers, American physicist
(1888–1968)
Meitner (11S, 113E): Lise Meitner, Austrian physicist
(1878–1968)
Mendel (49S, 110W): Gregor Johann Mendel, Austrian botanist
(1822–84)
Mendeleev (5N, 140E): Dmitri I. Mendeleev, Russian chemist
(1834–1907)

Menelaus (16N, 16E): Greek astronomer, first or second century AD

Mercator (29S, 26W): (Gerhard Kremer) Flemish geographer (1512–94)

Mercurius (56N, 65E): (Mercury) mythological messenger of gods (see planet *Mercury*)

Merrill (75N, 116W): Paul W. Merrill, American astronomer (1887–1961)

Mersenius (21S, 49W): (Marin Mersenne) French mathematician (1588–1648)

Mesentsev (72N, 129W): Yury B. Mezentsev, Soviet rocket engineer (1929–65)

Meshcerski (12N, 125E): Ivan V. Meshcerski, Russian engineer (1859–1935)

Messala (39N, 60E): (Mashalla) Jewish astronomer and astrologer, ninth century AD

Messier (2S, 48E): Charles Messier, French astronomer (1730–1817)

Metchnikoff (11S, 149W): Elie Metchnikoff (Ilya I. Mechnikov), Russian zoologist and bacteriologist, working in France (1845–1916)

Metius (40S, 44E): Adriaan Metius, Dutch mathematician and astronomer (1571–1635)

Meton (74N, 25E): Greek astronomer, fifth century BC

Michelson (6N, 121W): Albert A. Michelson, American physicist (1852–1931)

Milanković (77N, 170E): M. Milanković, Serbian mathematician (1879–1958)

Milichius (10N, 30W): (Jacob Milich) German physician and philosopher (1501–59)

Miller (39S, 1E): William Allen Miller, English chemist (1817–70)

Millikan (47N, 121E): Robert A. Millikan, American physicist (1868–1953)

Mills (9N, 156E): Mark M. Mills, American physicist (1917–58)

Milne (31S, 113E): Edward Arthur Milne, English astronomer (1896–1950)

Mineur (25N, 162W): Henri Mineur, French mathematician and astronomer (1899–1954)

Minkowski (56S, 145W): Hermann Minkowski, German mathematician (1864–1909)

Mitchell (50N, 20E): Maria Mitchell, American astronomer (1818–89)

Mitra (18N, 155W): S.K. Mitra, Indian physicist (1890–1963)

Möbius (16N, 101E): August Ferdinand Möbius, German mathematician and astronomer (1790–1868)

Mohorovičić (19S, 165W): Andrija Mohorovičić, Croatian geophysicist (1857–1936)

Moigno (66N, 28E): François Moigno, French mathematician (1804–84)

Moiseev (9N, 103E): Nikolai D. Moiseev, Soviet astronomer (1902–55)

Moltke (1S, 24E): Count Helmuth von Moltke, Prussian soldier (1800–91)

Monge (19S, 48E): Gaspard Monge, French mathematician (1746–1818)

Montgolfier (47N, 160W) (1) Jacques Etienne Montgolfier (1745–99);
                         (2) Joseph Michel Montgolfier (1740–1819), pioneer balloonists, brothers

Moore (37N, 178W): Joseph H. Moore, American astronomer (1878–1949)

Moretus (70S, 8W): Theodore Moretus, Flemish mathematician (1602–67)

Morozov (5N, 127E): Nikolai A. Morozov, Russian revolutionary scientist (1854–1946)

Morse (22N, 175W): Samuel F.B. Morse, American telegraph pioneer (1791–1872)

Mortis, Lacus (44N, 27E)

Moscoviense, Mare (25N, 147E)

Mösting (1S, 6W): Johann von Mösting, Danish statesman and astronomer (1759–1843)

Mouchez (86N, 35W): Amédée Ernest Barthélémy Mouchez, French astronomer (1821–92)

Murchison (5N, 0): Sir Roderick I. Murchison, Scottish-born geologist (1792–1871)

Mutus (63S, 30E): (Vincente Mut) Spanish astronomer (died 1673)

Nagoaka (20N, 154E): Hantaro Nagoaka, Japanese physicist (1865–1950)

Nansen (81N, 91E): Fridtjof Nansen, Norwegian explorer (1861–1930)

Nasireddin (41S, 0): (Nasir-ad-Din) Persian astronomer (1201–74)

Nasmyth (52S, 53W): James Nasmyth, Scottish engineer (1808–90)

Nassau (25S, 177E): Jason J. Nassau, American astronomer (1892–1965)

Naumann (35N, 62W): Karl Friedrich Naumann, German mineralogist (1797–1873)

Neander (31S, 40E): (Michael Neumann) German mathematician and astronomer (1529–81)

Nearch (58S, 39E): (Nearchus) Macedonian officer and commander, fourth century BC

Nebularum, Palus (38N, 1E)

Nectaris, Mare (14S, 34E)

Neison (68N, 28E): Edmund Neison, English astronomer (1849–1940)

Neper (7N, 83E): (John Napier) English mathematician (1550–1617)

Nernst (36N, 95W): Walther Hermann Nernst, German physicist and chemist (1864–1941)

Neujmin (27S, 125E): Grigori N. Neujmin, Soviet astronomer (1885–1946)

Neumayer (71S, 70E): Georg von Neumayer, German meteorologist (1826–1909)

Newcomb (30N, 44E): Simon Newcomb, American astronomer (1835–1909)

Newton (78S, 20W): Sir Isaac Newton, English mathematician and philosopher (1642–1727)

Nicholson (26S, 85W): Seth Barnes Nicholson, American astronomer (1891–1963)

Nicolaev (35N, 151E): Andrian G. Nicolaev, Soviet cosmonaut (born 1927)

Nicolai (42S, 26E): Friedrich Nicolai, German astronomer (1793–1846)

Nicollet (22S, 12W): Jean Nicolas Nicollet, French astronomer (1788–1843)

Niepce (72N, 120W): Joseph Nicéphore Niepce, French physicist (1765–1833)

Nijland (33N, 134E): Albertus A. Nijland, Dutch astronomer (1868–1936)

Nishina (45S, 171W): Yoshio Nishina, Japanese physicist (1890–1951)

Nobel (15N, 101W): Alfred Bernhard Nobel, Swedish inventor and philanthropist (1833–96)

Nöggerath (49S, 45W): Johann Jakob Nöggerath, German geologist (1788–1877)

Nonius (35S, 4E): (Pedro Nunes) Portuguese mathematician (1492–1577)

Nöther (66N, 114W): Emmy Nöther, German mathematician (1882–1935)

Nubium, Mare (19S, 12W)

Numerov (71S, 161W): Boris V. Numerov, Soviet astronomer (1891–1943)

Nušl (32N, 167E): František Nušl, Czech astronomer (1867–1925)

Obruchev (39S, 162E): Vladimir A. Obruchev, Russian geologist (1863–1956)

O'Day (31S, 157E): Marcus O'Day, American rocket-builder and physicist (1897–1961)

Oenopides (57N, 65W): Greek astronomer, fifth century BC

Oersted (43N, 47E): Hans Christian Oersted, Danish physicist (1777–1851)

Ohm (18N, 114W): Georg Simon Ohm, German physicist (1787–1854)

Oken (44S, 78E): Lorenz Oken, German naturalist and philosopher (1779–1851)

Olbers (7N, 78W): Heinrich Wilhelm Matthäus Olbers, German physician and astronomer (1758–1840) (see asteroids *Ceres*, *Pallas* and *Vesta*)

Olcott (20N, 117E): William T. Olcott, American writer on astronomy (1873–1936)

Omar Khayyám (58N, 102W): Persian poet and astronomer (died ?1123)

Opelt (16S, 18W): Friedrich Opelt, German financier and patron (1794–1863)

Oppenheimer (35S, 166W): Robert Oppenheimer, American physicist (1904–67)

Oppolzer (2S, 1W): Theodor Oppolzer, Austrian astronomer (1841–86)

Oresme (43S, 169E): Nicole Oresme, French prelate and scholar (?1330–82)

Orientale, Mare (19S, 95W)

Orlov (26S, 175W)　(1) Alexander Y. Orlov, Soviet astronomer (1880–1954);
(2) Sergei V. Orlov, Soviet astronomer (1880–1958)

Orontius (40S, 4W): (Oronce Fine) French mathematician (1494–1555)

Ostwald (11N, 122E): Wilhelm Ostwald, German physical chemist (1853–1932)

Palisa (9S, 7W): Johann Palisa, Austrian astronomer (1848–1925)

Pallas (5N, 2W): Peter Simon Pallas, German naturalist and traveller (1741–1811)

Palmieri (29S, 48W): Luigi Palmieri, Italian physicist (1807–96)

Paneth (63N, 95W): Friedrich Adolf Paneth, German chemist (1887–1958)

Pannekoek (4S, 140E): Anton Pannekoek, Dutch astronomer (1873–1960)

Paracelsus (23S, 163E): (Theophrastus Bombastus von Hohenheim) Swiss alchemist and physician (?1493–1541)

Paraskevopoulos (50N, 150W): John S. Paraskevopoulos, Greek astronomer (1889–1951)

Parenago (26N, 109W): Pavel P. Parenago, Soviet astronomer (1906–60)

Parkhurst (34S, 103E): John A. Parkhurst, American astronomer (1861–1925)

Parrot (15S, 3E): Johann Parrot, German physicist and explorer (1792–1840)

Parry (8S, 16W): Sir William Edward Parry, English explorer (1790–1855)

Parsons (37N, 171W): John W. Parsons, American rocket engineer (1913–52)

Pascal (74N, 70W): Blaise Pascal, French mathematician (1623–62)

Paschen (14S, 141W): Friedrich Paschen, German physicist (1865–1947)

Pasteur (11S, 105E): Louis Pasteur, French chemist (1822–95)

Pauli (45S, 137E): Wolfgang Pauli, Austrian physicist (1900–58)

Pavlov (29S, 142E): Ivan P. Pavlov, Russian physiologist (1849–1936)

Peary (89N, 50E): Robert E. Peary, American explorer (1856–1920)

Pease (13N, 106W): Francis G. Pease, American astronomer (1881–1938)

Peirce (18N, 53E): Benjamin Peirce, American mathematician (1809–88)

Peirescius (46S, 71E): (Nicolas de Peiresc) French astronomer
(1580–1637)

Pentland (64S, 12E): Joseph B. Pentland, Irish explorer
(1797–1873)

Perelman (24S, 106E): Yakov I. Perelman, Soviet rocket engineer
(1882–1942)

Perepelkin (10S, 128E): Yevgeny Y. Perepelkin, Soviet
astrophysicist (1906–40)

Petavius (25S, 61E): (Denys Petau) French theologian and scholar
(1583–1652)

Peterman (74N, 69E): August Petermann, German geographer
(1822–78)

Petropavlovsky (37N, 115W): Boris S. Petropavlovsky, Soviet
rocket designer (1898–1933)

Petrov (61S, 88E): Evgeny S. Petrov, Soviet rocket designer
(1900–42)

Pettit (28S, 86W): Edison Pettit, American astronomer
(1890–1962)

Petzval (63S, 113W): Joseph Mikas Petzval, Hungarian
mathematician (1807–91)

Phillips (26S, 78E): John Phillips, English geologist (1800–74)

Philolaus (75N, 33W): Greek philosopher, fifth century BC

Philolaus, Rupes (68N, 25W): 'Philolaus' Ridge', runs past crater
. Philolaus (above)

Phocylides (54S, 58W): Johannes Phocylides Holwarda (otherwise
Jan Fokker), Dutch astronomer (1618–51)

Piazzi (36S, 68W): Giuseppe Piazzi, Italian astronomer
(1746–1826) (see asteroids *Ceres* and *Piazzia*)

Piazzi Smyth (42N, 3W): Charles Piazzi Smyth, Scottish
astronomer (1819–1900) (see *Smythii, Mare*)

Picard (15N, 55E): Jean Picard, French astronomer (1620–82)

Piccolomini (30S, 32E): Alessandro Piccolomini, Italian prelate
and astronomer (1508–78)

Pickering (3S, 7E): Edward Charles Pickering, American
astronomer (1846–1919)

Pico, Mons (46N, 9W): 'Mount Pico', named by Schröter (see
below), who is said to have had 'Pico de Tenerife' ('Peak of
Tenerife') in mind (compare Piton)

Pictet (43S, 7W): Marc A. Pictet, Swiss astronomer (1752–1825)

Pirquet (20S, 140E): Baron Guido von Pirquet, Austrian space
scientist (1880–1966)

Pitatus (30S, 14W): (Pietro Pitati) Italian astronomer, sixteenth
century

Pitiscus (51S, 31E): Bartholomaeus Pitiscus, German
mathematician (1561–1613)

Piton, Mons (41N, 1W): 'Mount Piton', said to be named after a
peak in the Tenerife massif (word is French for 'peak');
compare Pico, Mons (above)

Pizzetti (35S, 119E): P. Pizzetti, Italian geodesist (1860–1918)

Plana (42N, 28E): Giovanni Plana, Italian astronomer
(1781–1864)

Planck (58S, 138E): Max Planck, German physicist (1858–1947)

Planck, Rima (65S, 129E; 54S, 125E): 'Planck's Rill', named after
crater Planck (see above) past which it runs

Plaskett (82N, 175E): John S. Plaskett, Canadian astronomer
(1865–1941)

Plato (51N, 9W): Greek philosopher (?427–347)

Playfair (23S, 9E): John Playfair, Scottish mathematician
(1748–1819)

Plinius (15N, 24E): (Pliny) Roman scholar (23–79)

Plummer (23S, 155W): Henry C.K. Plummer, English astronomer
(1875–1946)

Plutarch (25N, 75E): Greek biographer (?46–?120)

Pogson (42S, 111E): Norman Robert Pogson, English astronomer
(1829–91)

Poincaré (57S, 161E): Jules Henri Poincaré, French
mathematician (1854–1912)

Poinsot (79N, 147W): Louis Poinsot, French mathematician
(1777–1859)

Poisson (30S, 11E): Siméon Denis Poisson, French mathematician
(1781–1840)

Polybius (22S, 26E): Greek historian (?205–?125)

Polzunov (26N, 115E): Ivan I. Polzunov, Russian engineer
(1728–66)

Poncelet (76N, 55W): Jean Victor Poncelet, French
mathematician (1788–1867)

Pons (25S, 22E): Jean Louis Pons, French astronomer (1761–1831)

Pontanus (28S, 15E): (Giovanni Pontano) Italian humanist and
poet (1426–1503)

Pontécoulant (69S, 65E): Philippe Gustave Le Doulcet, Comte
de Pontécoulant, French mathematician and astronomer
(1795–1874)

Popov (17N, 100E): Alexander S. Popov, Russian radio pioneer (1859–1905)

Porter (56S, 10W): Russel W. Porter, American telescope-maker (1871–1949)

Posidonius (32N, 30E): Greek philosopher, first century BC

Poynting (17N, 133W): John Henry Poynting, English physicist (1852–1914)

Prager (4S, 131E): Richard A. Prager, German-born American astronomer (1884–1945)

Prandtl (60S, 141E): Ludwig Prandtl, German physicist (1875–1953)

Priestly (57S, 108E): Joseph Priestley, English chemist (1733–1804)

Prinz (26N, 44W): Wilhelm Prinz, German astronomer (1857–1910)

Procellarum, Oceanus (10N, 47W)

Proclus (16N, 47E): Greek philosopher (?410–185)

Proctor (46S, 5W): Mary Proctor, American astronomer (1862–1919)

Protagoras (56N, 7E): Greek philosopher, fifth century BC

Ptolemaeus (14S, 3W): (Ptolemy) Greek astronomer, second century AD

Puiseux (28S, 39W): Pierre Henri Puiseux, French astronomer (1855–1928)

Purbach (25S, 2W): Georg Purbach, Austrian mathematician and astronomer (1423–61)

Purkyně (1S, 95E): Jan Evangelista Purkyně, Czech physiologist (1787–1869)

Putredinis, Palus (27N, 1W)

Pyrenaei, Montes (14S, 41E): 'Pyrenees Mountains', named by Mädler (see above)

Pythagoras (65N, 65W): Greek philosopher and mathematician, sixth century BC

Pytheas (21N, 20W): Greek navigator and geographer, fourth century BC

Quételet (43N, 135W): Lambert Quételet, Belgian astronomer (1796–1874)

Rabbi Levi (35S, 24E): Levi ben Gershon, Spanish-Jewish philosopher (1288–1344)

Racah (14S, 180): Giulio Racah, Italian-Israeli physicist (1909–65)

Raimond (14N, 159W): J.J. Raimond, Dutch astronomer
(1903–61)

Ramsay (40S, 145E): Sir William Ramsay, Scottish chemist
(1852–1916)

Ramsden (33S, 32W): Jesse Ramsden, English astronomical
instrument-maker (1735–1800)

Rasumov (39N, 114W): Vladimir V. Razumov, Soviet rocket
engineer (1890–1967)

Rayet (45N, 114E): Georges Antoine Pons Rayet, French
astronomer (1839–1906)

Rayleigh (30N, 90E): John William Strutt Rayleigh, English
physicist (1842–1919)

Réaumur (2S, 1E): René Antoine Ferchault de Réaumur, French
naturalist and physicist (1683–1757)

Recta, Rupes (22S, 8W): 'Straight Fault' (more popularly 'Straight
Wall')

Regiomontanus (28S, 0): (Johann Müller) German astronomer
(1436–76)

Regnault (54N, 86W): Henri Victor Regnault, French chemist and
physicist (1810–78)

Reichenbach (30S, 48E): Georg von Reichenbach, German
astronomical instrument-maker (1772–1826)

Reimarus (46S, 55E): (Nicola Reymers Bär) German
mathematician (died 1600)

Reiner (7N, 55W): Vincentio Reinieri, Italian mathematician
(died 1648)

Reinhold (3N, 23W): Erasmus Reinhold, German mathematician
(1511–53)

Repsold (50N, 70W): Johann G. Repsold, German astronomical
instrument-maker (1771–1830)

Rhaeticus (0, 5E): (Georg Joachim von Lauchen) German
astronomer (1514–76)

Rheita (37S, 47E): Anton Maria Schyrleus of Rheita, Czech
optician (1597–1660)

Rheita, Vallis (40S, 48E): 'Rheita's Valley', on edge of which is
crater Rheita (see above)

Riccioli (3S, 75W): Giovanni Battista Riccioli, Italian philosopher
and astronomer (1598–1671)

Riccius (37S, 26E): (Matteo Ricci) Italian missionary in China
(1552–1610)

Ricco (75N, 177E): Annibale Ricco, Italian astronomer
(1844–1911)

Riedel (49S, 140W) (1) Klaus Riedel, German rocket engineer
(1907–44);
(2) Walter Riedel, German rocket engineer
(1902–68), brothers
Riphaeus, Montes (6S, 26W): 'Riphean Mountains', regarded by
Ancient Greeks as mountains from which north wind blew
Ritchey (11S, 9E): George Willis Ritchey, American astronomer
(1864–1945)
Rittenhouse (74S, 107E): David Rittenhouse, American
astronomer (1732–96)
Ritter (2N, 19E) (1) Karl Ritter, German geographer
(1779–1859);
(2) August Ritter, German astrophysicist
(1826–1908)
Ritz (15S, 92E): Walter Ritz, Swiss physicist (1878–1909)
Roberts (71N, 175W) (1) Alexander W. Roberts, South African
astronomer (1857–1938);
(2) Isaac Roberts, Welsh astronomer
(1829–1904)
Robertson (22N, 105W): Howard P. Robertson, American
physicist (1903–61)
Robinson (59N, 46W): Romney Robinson, Irish astronomer
(1792–1882)
Rocca (15S, 72W): Giovanni A. Rocca, Italian mathematician
(1607–56)
Roche (42S, 135E): Edouard Roche, French astronomer
(1820–83)
Römer (25N, 37E): Ole Roemer, Danish astronomer (1644–1710)
Röntgen (33N, 91W): Wilhelm Conrad Roentgen, German
physicist (1845–1923)
Rook, Montes (40S, 83W): 'Rook Mountains', named for
Lawrence Rooke, English astronomer (1622–66)
Roris, Sinus (54N, 46W)
Rosenberger (55S, 43E): Otto A. Rosenberger, German
mathematician (1800–90)
Ross (12N, 22E) (1) Sir James Clark Ross, Scottish explorer
(1800–62);
(2) Frank E. Ross, American astronomer
(1874–1966)
Rosse (18S, 35E): William Parsons, Earl of Rosse, Irish
astronomer (1800–67)

Rost (56S, 34W): Leonhardt Rost, German astronomer
(1688–1727)

Rothmann (31S, 28E): Christopher Rothmann, German
astronomer (died ?1600)

Rowland (57N, 163W): Henry A. Rowland, American physicist
(1848–1901)

Rozhdestvensky (86N, 155W): Dmitri S. Rozhdestvensky, Soviet
physicist and optician (1876–1940)

Rumford (29S, 170W): Benjamin Thompson, Count Rumford,
Anglo-American physicist (1753–1814)

Rümker, Mons (41N, 58W): 'Rümker's Mountain', named for
Karl Ludwig Christian Rümker, German astronomer
(1788–1862)

Russell (27N, 75W)   (1) John Russell, English painter and
astronomer (1745–1806);
(2) Henry Norris Russell, American
astronomer (1877–1957)

Rutherfurd (61S, 12W): Lewis M. Rutherfurd, American
astronomer (1816–92)

Rydberg (47S, 96W): Johannes Robert Rydberg, Swedish
physicist (1854–1919)

Rynin (37N, 86E): Nikolai A. Rynin, Soviet aeronautical engineer
(1877–1942)

Sabine (2N, 20E): Sir Edward Sabine, Irish astronomer
(1788–1883)

Sacrobosco (24S, 17E): (John of Holywood) English
mathematician, thirteenth century

Saenger (4N, 102E): Eugen Sänger, German aerospace engineer
(1905–64)

Šafařík (10N, 177E): Vojtech Šafařík, Czech astronomer
(1829–1902)

Saha (2S, 103E): Meghnad N. Saha, Indian physicist (1893–1956)

St John (10N, 150E): Charles E. St John, American astronomer
(1857–1935)

Sanford (32N, 139W): Roscoe F. Sanford, American astronomer
(1883–1958)

Santbech (21S, 44E): Daniel Santbech Noviomagus, Dutch
astronomer (died ?1561)

Sarton (49N, 121W): George A.L. Sarton, American science
scholar (1884–1956)

Sasserides (39S, 9W): (Gellio Sasceride) Danish physician (1562–1612)

Saunder (4S, 9E): Samuel A. Saunder, English astronomer (1852–1912)

Saussure (43S, 4W): Horace Bénédict de Saussure, Swiss geologist (1740–99)

Scaliger (27S, 109E): Joseph J. Scaliger, French-born German scholar (1540–1609)

Schaeberle (26S, 117E): John M. Schaeberle, American astronomer (1853–1924)

Scheiner (60S, 28W): Christoph Scheiner, German astronomer (?1579–1650)

Schiaparelli (23N, 59W): Giovanni Schiaparelli, Italian astronomer (1835–1910)

Schickard (44S, 54W): Wilhelm Schickard, German mathematician (1592–1635)

Schiller (52S, 39W): Julius Schiller, German monk and astronomer (died ?1627)

Schjellerup (69N, 157E): H.C. Schjellerup, Danish astronomer (1827–87)

Schlesinger (47N, 138W): Frank Schlesinger, American astronomer (1871–1943)

Schliemann (2S, 155E): Heinrich Schliemann, German archaeologist (1822–90)

Schlüter (7S, 84W): Heinrich Schlüter, German astronomer (1815–44)

Schmidt (1N, 19E) (1) Julius F.J. Schmidt, German astronomer (1825–84);
(2) Bernhard Schmidt, Estonian optician (1879–1935);
(3) Otto Y. Schmidt, Soviet explorer (1891–1956)

Schneller (42N, 146W): Herbert Schneller, German astronomer (1901–67)

Schömberger (76S, 30E): Georg Schoenberger, German mathematician (1597–1645)

Schönfeld (45N, 98W): Eduard Schönfeld, German astronomer (1828–91)

Schorr (19S, 90E): Richard Schorr, German astronomer (1867–1951)

Schrödinger (75S, 133E): Erwin Schrödinger, Austrian physicist (1887–1961)

Schrödinger, Rima (62S, 99E; 71S, 114E): 'Schrödinger's Rill', runs past crater Schrödinger (see above)

Schröter (3N, 7W): Johann Hieronymus Schröter, German astronomer (1745–1816)

Schröter, Vallis (26N, 52W): 'Schröter's Valley', named after above (but nowhere near crater Schröter)

Schubert (3N, 81E): Theodor F. von Schubert (Fëdor Fëdorovich Shubert), Russian geodesist (1789–1865)

Schumacher (42N, 60E): Heinrich Christian Schumacher, Danish astronomer (1780–1850)

Schuster (4N, 147E): Sir Arthur Schuster, British-Jewish physicist (1851–1934)

Schwarzschild (71N, 120E): Karl Schwarzschild, German astronomer (1873–1916)

Scoresby (80N, 25E): William Scoresby, English explorer (1789–1857)

Scott (82S, 45E): Robert Falcon Scott, English explorer (1868–1912)

Seares (74N, 145E): Frederick H. Seares, American astronomer (1873–1964)

Secchi (2N, 43E): Pietro Angelo Secchi, Italian astronomer (1818–78)

Sechenov (7S, 143W): Ivan M. Sechenov, Russian physiologist (1829–1905)

Segers (47N, 128E): Carlos Segers, Argentine astronomer (1900–67)

Segner (59S, 48W): Johann von Segner, German philosopher (1704–77)

Seidel (33S, 152E): Ludwig Philipp von Seidel, German physicist (1821–96)

Seleucus (21N, 66W): Babylonian astronomer, second century BC

Seneca (28N, 79E): Roman statesman and philosopher, first century AD

Serenitatis, Mare (30N, 17E)

Seyfert (29N, 114E): Carl Seyfert, American astronomer (1911–60)

Shajn (33N, 172E): Grigory A. Shajn, Soviet astronomer (1892–1956)

Shaler (33S, 88W): Nathaniel S. Shaler, American geologist (1841–1906)

Sharonov (13N, 173E): Vsevolod V. Sharonov, Soviet astronomer (1901–64)

Sharp (46N, 40W): Abraham Sharp, English mathematician (1651–1742)

Shatalov (24N, 140E): Vladimir A. Shatalov, Soviet cosmonaut (born 1927)

Sheepshanks (59N, 17E): Anne Sheepshanks, English benefactress (1789–1876)

Short (75S, 5W): James Short, Scottish mathematician and optician (1710–68)

Shuckburgh (43N, 53E): Sir George Shuckburgh, English astronomer (1751–1804)

Siedentopf (22N, 135E): H. Siedentopf, German astronomer (1906–63)

Sierpinski (27S, 155E): Waclaw Sierpinski, Polish mathematician (1882–1969)

Simpelius (75S, 15E): (Hugh Sempill) Scottish mathematician (1596–1654)

Sinas (9N, 32E): Simon Sinas, Greek patron (1810–76)

Sirsalis (13S, 60W): Girolamo Sirsalis, Italian astronomer (1584–1654)

Sirsalis, Rima (14S, 60W): 'Sirsalis' Rill', running past crater Sirsalis (above)

Sisakian (41N, 109E): Norair M. Sisakian, Soviet biochemist (1907–66)

Sklodowska (19S, 97E): Marya Skłodowska (Marie Curie), Polish physicist and chemist (1867–1934)

Slipher (50N, 160E)   (1) Earl C. Slipher, American astronomer (1883–1964);
                      (2) Vesto M. Slipher, American astronomer (1875–1969), brothers

Smoluchowski (60N, 96W): Marian Smoluchowski, Polish physicist (1872–1917)

Smythii, Mare (3S, 80E)

Snellius (29S, 56E): (Willebrord Snell) Dutch mathematician (1591–1626)

Sniadecki (22S, 169W): Jan Sniadecki, Polish astronomer (1756–1830)

Sommerfeld (65N, 161W): Arnold Sommerfeld, German physicist (1868–1951)

Sömmering (0, 7W): Samuel Thomas Sömmering, German naturalist (1755–1830)

Somnii, Palus (15N, 56E)

Somniorum, Lacus (37N, 35E)

Sosigenes (9N, 18E): Greek astronomer and mathematician, first century BC

Spallanzani (46S, 25E): Lazzaro Spallanzani, Italian naturalist (1729–99)

Spencer Jones (13N, 166E): Sir Harold Spencer Jones, English astronomer (1890–1960)

Spörer (4S, 2W): Gustav Friedrich Wilhelm Spörer, German astronomer (1822–95)

Spumans, Mare (1N, 65E)

Stadius (11N, 14W): (Jan Stade) Belgian astronomer (1527–79)

Stark (25S, 134E): Johannes Stark, German physicist (1874–1957)

Stebbins (65N, 143W). Joel Stebbins, American astronomer (1878–1966)

Stefan (46N, 109W): Josef Stefan, Austrian physicist (1835–93)

Stein (7N, 179E): J.W. Stein, Dutch astronomer (1871–1951)

Steinheil (50S, 48E): Carl August von Steinheil, German physicist (1801–70)

Steklov (37S, 105W): Vladimir A. Steklov, Soviet mathematician (1863–1926)

Steno (33N, 162E): Nicolaus Steno, Danish physician (1638–86)

Šternberg (19N, 117W): Pavel K. Šternberg, Soviet astronomer (1865–1920)

Stetson (40S, 119W): Harlan T. Stetson, American astronomer (1885–1964)

Stevinus (33S, 54E): (Simon Stevin) Dutch mathematician (1548–1620)

Stiborius (34S, 32E): (Andreas Stoberl) Austrian philosopher (1465–1515)

Stöfler (41S, 6E): Johan Stöffler, German mathematician (1452–1534)

Stokes (54N, 87W): Sir George Gabriel Stokes, Irish mathematician (1819–1903)

Stoletov (45N, 155W): Alexander G. Stoletov, Russian physicist (1839–96)

Stoney (56S, 156W): George Johnstone Stoney, Irish physicist (1826–1911)

Størmer (57N, 145E): Fredrick C.M. Størmer, Norwegian mathematician (1874–1957)

Strabo (62N, 55E): Greek geographer, first century BC and AD

Stratton (6S, 165E): Frederick J.M. Stratton, English astronomer (1881–1960)

Street (46S, 10W): Thomas Street, English astronomer (died
?1661)
Strömgren (22S, 133W): Elis Strömgren, Swedish astronomer
(1870–1947)
Struve (43N, 65E) (1) Friedrich G.W. von Struve, German
astronomer (1793–1864);
(2) Otto Wilhelm von Struve, Russo-German
astronomer (1819–1905);
(3) Otto Struve, Russo-American
astronomer (1897–1963), father, son
and great-grandson
Subbotin (29S, 135E): Mikhail F. Subbotin, Soviet astronomer
(1893–1966)
Sulpicius Gallus (20N, 12E): Roman consul and scholar, second
century BC
Sumner (37N, 109E): Thomas H. Sumner, American geographer
(1807–76)
Sylvester (83N, 80E): James Joseph Sylvester, English
mathematician (1814–97)
Szilard (34N, 106E): Leo Szilard, American physicist (1898–1964)

Tacitus (16S, 19E): Cornelius Tacitus, Roman orator and historian
(?55–?120)
Tacquet (17N, 19E): André Tacquet, Belgian mathematician
(1612–60)
Tannerus (56S, 22E): (Adam Tanner) German mathematician
(1572–1632)
Taruntius (6N, 46E): Lucius Taruntius Firmanus, Roman
astrologer, first century BC
Taurus, Montes (28N, 35E): 'Taurus Mountains', name
(translating as 'Bull Mountains') given by Hevelius, on
uncertain reasoning
Taylor (5S, 17E): Brook Taylor, English mathematician
(1685–1731)
Teisserenc (32N, 137W): Léon Philippe Teisserenc de Bort,
French meteorologist (1855–1913)
Tempel (4N, 12E): Ernst Wilhelm Leberecht Tempel, German
astronomer (1821–89)
Ten Bruggencate (9S, 134E): P. Ten Bruggencate, German
astronomer (1901–61)
Tesla (38N, 125E): Nikola Tesla, Yugoslav electrician and
inventor (1857–1943)

Thales (59N, 41E): Greek philosopher and scientist (?640–546)

Theaetetus (37N, 6E): Greek philosopher, fourth century BC

Thebit (22S, 4W): Thebit ben Korra, Arab astronomer (826–901)

Theon (A) (1S, 15E): Theon of Smyrna, Greek astronomer, first or second century AD

Theon (B) (2S, 16E): Theon of Alexandria, Greek astronomer, fourth century AD

Theophilus (12S, 26E): (St Theophilus) Egyptian theologian, fifth century AD

Thiel (40N, 134W): Walter Thiel, German rocket engineer (1910–43)

Thiessen (75N, 169W): E. Thiessen, German astronomer (1914–61)

Thomson (32S, 166E): Sir Joseph J. Thomson, English physicist (1856–1940)

Tikhov (62N, 172E): Gavriil A. Tikhov, Soviet astronomer (1875–1960)

Tiling (52S, 132E): Reinhold Tiling, German rocket engineer (1890–1933)

Timaeus (63N, 1W): Greek philosopher, fifth century BC

Timiryazev (5S, 147W): Kliment A. Timiryazev, Russian naturalist (1843–1920)

Timocharis (27N, 13W): Greek astronomer, third century BC

Tisserand (21N, 48E): François Félix Tisserand, French astronomer (1845–96)

Titius (27S, 101E): Johann D. Titius, German mathematician (1729–96)

Titov (28N, 150E): German S. Titov, Soviet cosmonaut (born 1935)

Torricelli (5S, 29E): Evangelista Torricelli, Italian mathematician (1608–47)

Tralles (28N, 53E): Hohann G. Tralles, German physicist (1763–1822)

Tranquillitatis, Mare (9N, 30E)

Triesnecker (4N, 4E): Franz von Paula Triesnecker, Austrian astronomer (1745–1817)

Trümpler (28N, 168E): Robert J. Trumpler, Swiss astronomer (1886–1956)

Tsander (5N, 149W): Fridrikh A. Tsander, Soviet rocket pioneer (1887–1933)

Tsiolkovsky (21S, 128E): Konstantin E. Tsiolkovsky, Soviet space flight pioneer (1857–1935)

Turner (2S, 13W): Herbert Hall Turner, English astronomer (1861–1930)
Tycho (43S, 11W): Tycho Brahe, Danish astronomer (1546–1601)
Tyndall (35S, 117E): John Tyndall, Irish physicist (1820–93)

Ukert (8N, 1E): Friedrich August Ukert, German historian and philologist (1780–1851)
Ulugh Beigh (29N, 89W): Uzbek astronomer and mathematician (1394–1449)

Valier (7N, 174E): Max Valier, German engineer and rocket pioneer (1895–1930)
Van de Graaff (27S, 172E): Robert Jemison Van de Graaff, American physicist (1901–67)
Van den Bergh (31N, 159W): G. Van den Bergh, Dutch lawyer and mathematician (1890–1966)
Van der Waals (44S, 119E): Johannes Diderk van der Waals, Dutch physicist (1837–1923)
Van Gent (16N, 160E): H. Van Gent, Dutch astronomer (1900–47)
Van Rhijn (52N, 145E): Pieter J. Van Rhijn, Dutch astronomer (1886–1960)
Van't Hoff (62N, 133W): Jacobus Hendricus van't Hoff, Dutch physical chemist (1852–1911)
Vaporum, Mare (14N, 5E)
Vasco da Gama (15N, 85W): Portuguese navigator (1469–1524)
Vashakidze (44N, 93E): Mikhail A. Vashakidze, Soviet astronomer (1909–56)
Vavilov (1S, 139W) (1) Nikolai I. Vavilov, Soviet botanist (1887–1943);
              (2) Sergei I. Vavilov, Soviet physicist and optician (1891–1951), brothers
Vega (45S, 63E): Georg, Freiherr von Vega, German mathematician (1756–1802)
Vendelinus (16S, 62E): (Godefroid Wendelin) Flemish astronomer (1580–1667)
Vening Meinesz (0, 163E): Felix A. Vening Meinesz, Dutch geodesist (1887–1966)
Ventris (5S, 158E): Michael G.F. Ventris, British scholar (1922–56)
Vernadsky (23N, 130E): Vladimir I. Vernadsky, Soviet geochemist (1863–1945)

Verne, Jules (38S, 145E): French writer (1828–1905)

Vesalius (3S, 115E): (Andreas Vesele) Flemish physician (1514–64)

Vestine (34N, 94E): Ernest H. Vestine, American geophysicist (1906–68)

Vetchinkin (10N, 131E): Vladimir P. Vetchinkin, Soviet aerodynamics engineer (1888–1950)

Vieta (29S, 57W): (François Viète) French mathematician (1540–1603)

Vil'ev (6S, 144E): Mikhail A. Vil'ev, Russian astronomer (1893–1919)

Vitello (30S, 38W): (Erazm Ciołek) Polish mathematician (?1210–?1285)

Vitruvius (18N, 31E): Pollio Vitruvius, Roman architect, first century BC

Vlacq (53S, 39E): Adrian Vlacq, Dutch mathematician (?1600–67)

Vogel (15S, 6E): Hermann Karl Vogel, German astrophysicist (1841–1907)

Volta (53N, 85W): Count Alessandro Volta, Italian physicist (1745–1827)

Volterra (57N, 131E): Vito Volterra, Italian physicist (1860–1940)

Von der Pahlen (25S, 133W): Emanuel Von der Pahlen, German astronomer (1882–1952)

Von Kármán (45S, 176E): Theodor von Kármán, Hungarian physicist (1881–1963)

Von Neumann (40N, 153E): John von Neumann, American mathematician (1903–57)

Von Zeipel (42N, 142W): E.H. Von Zeipel, Swedish astronomer (1873–1959)

Walker (26S, 162W): Joseph A. Walker, American test pilot (1921–66)

Wallace (20N, 9W): Alfred Russel Wallace, English naturalist (1823–1913)

Walter (33S, 1E): Bernard Walter, German astronomer (1430–1504)

Wargentin (59S, 60W): Pehr Vilhelm Wargentin, Swedish astronomer (1717–83)

Waterman (26S, 128E): Alan T. Waterman, American physicist (1892–1967)

Watson (63S, 124W): James Craig Watson, Canadian-born American astronomer (1838–80)

Watt (50S, 51E): James Watt, Scottish engineer (1736–1819)

Webb (1S, 60E): Thomas William Webb, English astronomer (1806–85)

Weber (50N, 124W): Wilhelm Eduard Weber, German physicist (1804–91)

Wegener (45N, 113W): Alfred Lothar Wegener, German geophysicist (1880–1930)

Weinek (28S, 37E): Ladislav Weinek, Austrian astronomer (1848–1913)

Weiss (32S, 20W): Edmund Weiss, Austrian astronomer (1837–1917)

Wells, H.G. (41N, 122E): English novelist (1866–1946)

Werner (28S, 3E): Johan Werner, German astronomer (1468–1528)

Wexler (69S, 90E): Harry Wexler, American meteorologist (1911–62)

Weyl (16N, 120W): Hermann Weyl, German-born American mathematician (1885–1955)

Whewell (4N, 14E): William Whewell, English philosopher (1794–1866)

White (48S, 149W): Edward H. White, American astronaut (1930–67)

Wichmann (8S, 38W): Moritz Wichmann, German astronomer (1821–59)

Wiener (41N, 146E): Norbert Wiener, American mathematician (1894–1964)

Wilhelm (43S, 20W): (William IV of Hesse) German statesman and astronomer (1532–92)

Wilkins (30S, 20E): Hugh Percival ('Percy') Wilkins, Welsh astronomer (1896–1960)

Williams (42N, 37E): Arthur Stanley Williams, English lawyer and astronomer (1861–1938)

Wilsing (22S, 155W): J. Wilsing, German astronomer (1856–1943)

Wilson (69S, 33W) (1) Alexander Wilson, Scottish astronomer (1714–86);
(2) Charles T.R. Wilson, Scottish physicist (1869–1959);
(3) Ralph E. Wilson, American astronomer (1866–1960)

Winkler (42N, 179W): Johannes Winkler, German rocket
engineer (1897–1947)
Winlock (35N, 106W): Joseph Winlock, American astronomer
(1826–75)
Wöhler (38S, 31E): Friedrich Wöhler, German chemist (1800–82)
Wolf (23S, 17W): Max J.C. Wolf, German astronomer
(1863–1932)
Wollaston (31N, 47W): William Hyde Wollaston, English chemist
and physicist (1766–1828)
Woltjer (45N, 160W): Jan Woltjer, Dutch astronomer
(1891–1946)
Wood (44N, 121W): Robert Williams Wood, American physicist
(1868–1955)
Wright (31S, 88W) (1) Frederick E. Wright, American
astronomer (1878–1953);
(2) Thomas Wright, English philosopher
(1711–86);
(3) William H. Wright, American
astronomer (1871–1959)
Wrottesley (24S, 57E): John, Baron Wrottesley, English
astronomer (1798–1867)
Wurzelbauer (34S, 16W): Johann Philipp Wurzelbauer, German
astronomer (1651–1725)
Wyld (1S, 98E): James H. Wyld, American rocket pioneer
(1913–53)

Xenophenes (56N, 80W): Greek philosopher, sixth century BC

Yamamoto (59N, 161E): I. Yamamoto, Japanese astronomer
(1889–1959)
Yerkes (15N, 52E): Charles T. Yerkes, American financier
(1837–1905)
Young (42S, 51E): Thomas Young, English physician and
physicist (1773–1829)

Zach (61S, 5E): Franz Xaver, Freiherr von Zach, Hungarian
astronomer (1754–1832)
Zagut (32S, 22E): Abraham ben Samuel Zaguth, Spanish-Jewish
astronomer, fifteenth century
Zeeman (75S, 135W): Pieter Zeeman, Dutch physicist
(1865–1943)
Zeno (46N, 75E): Greek philosopher, fourth and third century BC

Zernike (18N, 168E): Frits Zernike, Dutch physicist (1886–1966)

Zhiritsky (25S, 120E): Georgy S. Zhiritsky, Soviet aeronautical engineer (1893–1966)

Zhukovsky (7N, 167W): Nikolai Y. Zhukovsky, Russian aerodynamics pioneer (1847–1921)

Zinger (57N, 176E): Nikolai Y. Zinger, Russian astronomer (1842–1918)

Zöllner (8S, 19E): Johann Karl Friedrich Zöllner, German astrophysicist (1834–82)

Zsigmondy (59N, 105W): Richard Zsigmondy, German chemist (1865–1929)

Zucchius (61S, 50W): (Niccolo Zucchi) Italian mathematician (1586–1670)

Zupus (17S, 52W): (Giovanni B. Zupi) Italian astronomer (?1590–1650)

APPENDIX II: NAMES OF MINOR PLANETS

The list that follows gives brief details of the first 1000 minor planets (asteroids) to be discovered and named, with the number that was officially allocated to the individual planet. (Names in *italics* have their own entry in the Dictionary.)

Two things catch the eye immediately, as one glances down the list. The first is that the majority of names are feminine, and the second is that most of them end in '-a' or '-ia'.

For the early names, classical mythology plays an important role, and many familiar names from the stories of classical mythology will be recognised, together with less familiar names. The first ten names, too, are those of Ancient Greek or Roman goddesses. But with planet 12 we already find an early deviation from the classical theme, as the name is that of a human monarch, moreover a living one at the time of naming.

As the names progress, one also finds a noticeable sprinkling of 'personified abstractions', names that may have been those of mythological characters, but that likewise serve as the embodiments of some quality or attribute, such as peace (58), justice (99) or honour (236). Some of these may have been motivated by a historic event, for example 58 Concordia and 306 Unitas. It is not uncommon to find Greek and Roman counterparts here, as it is for the names of mythological characters, so that one classical pair is 4 *Vesta* and 46 Hestia, another is 78 Diana and 105 Artemis, and a third is 227 Philosophia (Greek) and 275 Sapientia (Latin).

Two other types of name then appear, even before the first hundred have been completed. Ancient geographical names are introduced (such as 51 Nemausa and 52 *Europa*, the latter doubling as a mythological name) and the mythological theme broadens with names of Norse and other gods and goddesses (such as 76 Freia and 77 Frigga).

Fairly early on, too, the names of members of a discoverer's family are used as personal tributes, a theme that is pursued increasingly from about planet 300. Examples prior to this are 154 Bertha and 169 Zelia. Both the latter names were given by the French astronomer Flammarion, who was a great 'family namer'. In planet 107 Camilla he even records his own name, Camille, but with the by now traditional feminine ending of '-a'. This principle is important, because it means that many apparently feminine names will actually be of male personages, whether mythological or real, for example 54 *Alexandra*, named for an Alexander.

After planet 100 (a true 'centenary' name, as will be seen), other sources of names are increasingly utilised, such as literary characters (152 Atala, 171 Ophelia, 282 Clorinde) and (from about planet 500) musical works, especially operas (528 Rezia, 531 Zerlina, 539 Pamina, 550 Senta).

Of the 'commemorative' names, it will be noted that many refer to the place of discovery, as well as the name of the discoverer. A rare early example is 64 Angelina, named for the French observatory where it was first observed; other such names are 142 Polana, 263 Dresda, 325 Heidelberga, 384 Burdigala. (A more general changing pattern can be interestingly discerned in the dominant nationality of the names at any point. Most of the early names are of French creation. German names then become increasingly common, mingled with Italian, until about planet 700, when first American, then Russian names are much in evidence. In a sense, the names are a sort of potted history of astronomical discovery, much as geographical place-names are or were.)

By the time one has reached planet 88, mythological names are a rarity, and only very occasionally does one find the name of a mythical character introduced. This was partly because many of the obvious names had simply been used up, but it was also because the naming became much more obviously and deliberately 'meaningful'. At the same time, it has to be admitted that many quite recent names (including those of the second and third thousand not listed here) are frustratingly anonymous. We still do not know, for instance, who the original bearers were of names such as 821 Fanny, 842 Kerstin, 874 Rotraut, 939 Isberga. Such names *may* be tributes to historical or mythical (or fictional) people, but equally they may simply be personal commemorations of people known only to the namers. One can assume, with a fifty-fifty probability, that 985 Rosina and 986 Amelia were women or girls known to their namer (daughter, wife, fiancée, friend), but they could merely be random names. One can only go by what has actually been discovered about the sources of the names; it is known that some discoverers simply invented feminine-sounding names. See, for example, 913 Otilia, 918 Itha and many succeeding ones in this group.

This brings us back to the matter of the gender of the names. While most names are predominantly feminine, some are noticeably masculine. From about the year 1900, masculine names were deliberately selected for those minor planets whose orbits were different from the norm. Most minor planets have orbits that lie between Mars and Jupiter. Some, however, deviate from this general rule, and the so-called Trojans, for example, move in virtually the same orbit as Jupiter, and lie both east and west of this planet. Because of their different orbits, they therefore have masculine names (such as 588 *Achilles*, 617 *Patroclus* and 624 *Hector*), and names that, moreover, link them mythologically with the Trojan War (see *Trojans* in Dictionary). The first minor planet of this group to be named was *Achilles*, but this was not the first masculine name to be given. That distinction belongs to 342 Endymion, which lacks the final feminine '-a'. (Although before this, names of males can be seen behind other names, such as 54 *Alexandra* already mentioned.)

Endymion was given a masculine name because the namers had (temporarily) run out of feminine mythological ones. As a minor planet, however, Endymion has no orbital or other peculiarity. Later discoveries, that *did* have such peculiarities and so were deliberately given masculine names, were 433 *Eros* (the first to be so assigned for this reason, in 1898), 1221 *Amor* and 1566 *Icarus*. *Amor* has its orbit, like that of *Eros* (of whose name it is the Latin counterpart) between Mars and Venus, not Mars and Jupiter, while *Icarus* is distinctive in being the only minor planet so far discovered to have an orbit between Mercury and the Sun. The classical symbolism of the name is obvious. (But if it is not, see the entry in the Dictionary.) Similarly *Hermes* (which has no number) has approached close to the Earth, and the orbit of 1862 *Apollo* crosses that of the Earth.

The use of name gender and endings to denote astronomical information like this is very similar to the '-e' ending used for four of Jupiter's satellites (see the Introduction, p. 22, and also *Pasiphaë*).

Appropriately, minor planet 1000 *Piazzia* has a name of special 'thousand up' significance, as it commemorates the discoverer of planet 1 *Ceres*. The special '-ia' ending on his name, needed to make it conform to the 'feminine' principle, can also be seen on many other names of males as well as on place-names, for example in 727 Nipponia, 729 Watsonia, 749 Malzovia and 757 Portlandia. The use of this suffix for scientific denotation is nothing new, and many familiar flower names consist of a (male) surname with the same '-ia', for example 'aubrietia' (named after Aubriet), 'dahlia' (after Dahl) and 'lobelia' (after Lobelius, otherwise Mathias de l'Obel). Flower names themselves sometimes also serve as minor planet names, although in cases where they coincide with feminine names (e.g. 957 Camelia or 970 Primula) one cannot be certain whether it is a plant or a person.

After 1000 *Piazzia* names continued in much the same vein as they had now reached, with one or two forays into the experimental, however, which should perhaps be mentioned here for the sake of completeness.

Minor planet 1019 was named Strackea, for the German astronomer Gustav Stracke, and his surname was again commemorated in minor planets 1227 Geranium, 1228 Scabiosa, 1229 Tilia, 1230 Riceia, 1231 Auricula, 1232 Cortusa, 1233 Kobresia and 1234 Elyna. The initials of these eight names spell out G. STRACKE, and this unusual method of naming a colleague was devised by the discoverer of the eight planets, Karl Reinmuth. (For more of Reinmuth's naming devices, see 913 Otilia in the list below.)

Similar word play was devised by the veteran (and venturous) namer Reinmuth for 1572 Haremari, which was intended as a light-hearted tribute to his female colleagues, his 'harem'! (He could get away with that in 1935. . .)

Other names include those of a Mozambiquan city (1474 Beira), an African tribe and its language (1506 Xhosa), the nickname (meaning 'shirtless woman') adopted by Eva Perón (1588 Descamisada), a Soviet cosmonaut (1772 Gagarin), one of King Arthur's knights (2082 Galahad), an airline (2138 Swissair), a footballer (2202 Pele), a character in a popular television science fiction/space adventure (2309 Mr Spock), an American painter (2529 Rockwell Kent), a Russian poet (2576 Yesenin) and an English city (3009 Coventry). (Chaucer makes it for minor planet 2984, and Shakespeare follows him for 2985.)

As for 1000 *Piazzia*, special names were selected for 2000 Herschel, to honour the German-born English astronomer Sir William Herschel and his family (his son John and his sister Caroline were also noted astronomers), and for 3000 Leonardo, to pay tribute to the versatility and genius of Leonardo da Vinci. This great Italian is accompanied by his famous fellow countrymen 2999 Dante and 3001 Michelangelo.

In early 1987 the latest minor planet to have been named was 3288 Seleucus, honouring one of the grand generals of Alexander the Great.

Names of minor planets that have been lost since their original discovery are 719 Albert, 724 Hapag and 878 Mildred (see the list below for the origins of these names, if known); 1179 Mally, discovered in 1931, was also lost for many years but was rediscovered in late 1986. ('Presumably Mally was the name of his girl-friend', wrote *The Times* in its issue of 1 January 1987, referring to the discoverer, the famous Reinmuth. But it was actually the name of a daughter-in-law of the astronomer Max Wolf.)

The following minor planets with numbers above 1000 have their own entries in the Dictionary: 1172 *Aeneas*, 1221 *Amor*, 1862 *Apollo*, 2060 *Chiron*, 1665 *Gabi*, 1001 *Gaussia*, 1620 *Geographos*, 1566 *Icarus*, 1486 *Marilyn*, 1086 *Nata* and 1625 *The NORC*.

1 *Ceres*
2 *Pallas* (compare 93 and 881)
3 *Juno*
4 *Vesta*
5 *Astraea*
6 *Hebe*
7 *Iris*
8 *Flora*
9 *Metis*
10 *Hygeia*
11 *Parthenope*
12 *Victoria*
13 Egeria: obscure Roman nymph
14 Irene: Greek goddess of peace
15 Eunomia: Greek goddess of order
16 Psyche: beautiful Greek girl, personification of the soul
17 Thetis: one of Nereids, mother of Achilles
18 Melpomene: Greek muse of tragedy (see 600)
19 Fortuna: Roman goddess of fortune
20 Massalia: Latin name of Marseille, where namer, French
    astronomer Chacornac, worked
21 Lutetia: Latin name of Paris; planet also named by French
22 Calliope: Greek muse of epic poetry (see 600)
23 Thalia: Greek muse of comedy and pastoral poetry (see 600)
24 *Themis*
25 Phocaea: ancient Ionian port in Asia Minor

26 Proserpina: Roman goddess of underworld (Greek Persephone; see 399)

27 Euterpe: Greek muse of lyric poetry and music (see 600)

28 Bellona: Roman goddess of war

29 Amphitrite: Greek sea goddess, wife of Poseidon

30 Urania: Greek muse of astronomy (see 600)

31 Euphrosyne: one of three Graces; name means 'cheerfulness' (compare 39)

32 Pomona: Roman goddess of fruit trees

33 Polyhymnia: Greek muse of singing, mime and sacred dance (see 600)

34 Circe: Greek sorceress who detained Odysseus on her island

35 Leucothea: Greek goddess into whom Ino, daughter of Cadmus, was turned (see 173)

36 Atalante: Greek virgin huntress

37 Fides: Roman goddess of good faith

38 Leda: Greek Queen of Sparta, visited by Zeus disguised as a swan

39 Laetitia: Latin for 'gladness' (compare 31)

40 Harmonia: Greek daughter of Ares (Mars) and Aphrodite (Venus)

41 Daphne: Greek nymph pursued by Apollo

42 Isis: ancient Egyptian fertility goddess (compare 161)

43 Ariadne: Greek daughter of Minos who helped Theseus kill Minotaur

44 *Nysa*

45 Eugenia: named for Eugénie, French empress, wife of Napoleon III

46 Hestia: Greek goddess of the hearth (Roman *Vesta*)

47 Aglaia: one of three Graces

48 *Doris*

49 Pales: Roman god of flocks and shepherds

50 Virginia: named for American state of Virginia

51 Nemausa: Latin name of Nîmes, French town where astronomers discovered planet

52 *Europa*

53 Calypso: Greek goddess, daughter of Atlas

54 *Alexandra*

55 Pandora: first Greek woman, who released ills of man from 'Pandora's box'

56 Melete: Greek for 'care', 'anxiety'; planet proved difficult to identify

57 Mnemosyne: Greek goddess of memory, mother of the Muses (see 600)

58 Concordia: Latin for 'peace', symbolic of end of Franco-Italo-Austrian war (1859; compare 679)

59 Elpis: Greek for 'hope'; planet was long unnamed, so good name was hoped for

60 Echo: Greek nymph who repeated words of others

61 Danaë: Greek mother of Perseus

62 Erato: Greek muse of love poetry (see 600)

63 Ausonia: old poetic name of Italy; planet was discovered (1861) by De Gasparis

64 Angelina: named for observatory at Notre Dame des Anges, near Marseille (see 999)

65 Cybele: Phrygian goddess of nature

66 Maia: one of Pleiades, mother by Zeus (Jupiter) of Hermes (Mercury)

67 Asia: named for continent (like *Europa*), with Asia also mother of Atlas

68 Leto: Greek daughter of Titans (Roman Latona; see 639)

69 Hesperia: Greek poetic name for Italy; discovered by Schiaparelli (1861)

70 Panopaea: one of Nereids (compare 17), daughters of *Doris*

71 Niobe: first mortal woman loved by Zeus (Jupiter)

72 Feronia: Roman goddess of freed slaves; named at start of American Civil War

73 Clytia: young Greek girl loved by Helios, the Sun

74 Galatea: one of Nereids, a 'milk-white maiden' (compare *Galaxy* in Glossary)

75 Eurydice: Greek nymph who married Orpheus

76 Freia: Norse goddess of love and fertility (Roman Venus; compare 77)

77 Frigga: Norse goddess of marriage, wife of Odin (Roman *Juno*)

78 Diana: Roman goddess of the hunt and the Moon (Greek Artemis; compare 105 and 394)

79 Eurynome: one of Oceanids, mother of the Graces

80 Sappho: named for Greek poetess of sixth century BC

81 Terpsichore: Greek muse of dance and choral song (see 600)

82 Alcmene: mother of Hercules by Jupiter

83 Beatrix: named either for Florentine noblewoman or discoverer's daughter

84 Clio: Greek muse of history (see 600)

85 Io: maiden loved by Zeus (Jupiter) and turned into white
    heifer by him

86 Semele: mother of Dionysus (Bacchus) by Zeus (Jupiter)

87 Sylvia: named for Sylvie, first wife of French astronomer
    Flammarion

88 Thisbe: tragic lover of Pyramus (see Shakespeare's
    A Midsummer Night's Dream)

89 Julia: unidentified; perhaps member of discoverer's family
    (Pogson, 1866)

90 Antiope: daughter of Asopus (or Nycteus) loved by Zeus
    (Jupiter; compare 91)

91 Aegina: daughter of river god Asopus (compare 90)

92 *Undina*

93 Minerva: Roman goddess of wisdom (Greek Athena;
    compare 2 and 881)

94 Aurora: Roman goddess of the dawn (Greek Eos; compare
    221 and 700)

95 Arethusa: nymph changed into a spring to escape advances
    of Alpheus

96 Aegle: daughter of Panopeus loved by Theseus

97 Clotho: one of three Fates, spinner of thread of life (see 120
    and 273)

98 Ianthe: one of Oceanids (daughters of Oceanus and Tethys)

99 Dike: goddess personifying justice, together with *Themis*

100 Hecate: goddess of the underworld (Greek *hekaton* means
    'hundred')

101 Helena: 'Helen of Troy', the beautiful daughter of Zeus
    (Mercury) and Leda

102 Miriam: sister of Moses and Aaron (biblical name)

103 Hera: Queen of Olympian gods, sister and wife of Zeus
    (Roman Juno; compare 3)

104 Clymene: Oceanid, mother of Atlas by Iapetus

105 Artemis: goddess of the hunt (Roman Diana; compare 78)

106 Dione: daughter of Uranus and Ge (or of Oceanus and
    Tethys, or of Atlas)

107 Camilla: first name (in feminine form) of French astronomer
    Flammarion

108 Hecuba: wife of King Priam of Troy, mother of Hector and
    Paris

109 Felicitas: Latin for 'happiness', name of goddess
    personifying this

110 Lydia: probably named for ancient region of Asia Minor, or for a woman/girl

111 Ate: goddess who blinded men so that they erred into guilty deeds

112 Iphigenia: daughter of Agamemnon, saved from sacrifice by Artemis

113 Amalthea: nymph who brought up baby Zeus (Jupiter) on goat's milk

114 Cassandra: prophetess, daughter of Priam and Hecuba

115 Thyra: name of early Danish queen; discoverer (Watson) had been to Denmark

116 Sirona: Celtic (Gaulish) goddess worshipped at warm springs

117 Lomia: unidentified; perhaps private reference to discoverer's family

118 Peitho: goddess of persuasion, daughter of Prometheus (in some accounts)

119 Althaea: mother of Meleager by her husband (her uncle) Oeneus

120 Lachesis: one of three Fates, with name derived from Greek *lakhesis*, 'destiny' (see 97 and 273)

121 Hermione: daughter of Menelaus and Helen

122 Gerda: in Scandinavian mythology, goddess of the frozen Earth, wife of Frey

123 Brunhild: in Scandinavian mythology, one of Valkyries, daughters of Odin

124 Alceste: wife of Alcestis of Thessaly, saved from Hades by Hercules

125 Liberatrix: Latin for 'liberator woman', perhaps named symbolically for liberation of France after Franco-Prussian War (1871), or named for Joan of Arc

126 Velleda: priestess worshipped by Germans in Roman times (mentioned by Tacitus)

127 *Johanna*

128 Nemesis: goddess of retribution and vengeance

129 Antigone: daughter of Oedipus and Jocasta

130 Electra: daughter of Agamemnon and Clytemnestra

131 Vala: in Indian mythology, god of caves

132 Aethra: mother of Theseus (planet was lost for forty-nine years until 1922)

133 Cyrene: nymph loved by Apollo

134 Sophrosyne: Greek for 'prudence', one of four virtues in Plato's system (compare Latin form, 474)

135 Hertha: Germanic goddess of the earth or of fertility
136 Austria: named for country, as first minor planet to be
    discovered there
137 Meliboea: daughter of Oceanus; also name of ancient town
    in Thessaly
138 Tolosa: Latin name of Toulouse, France, where discovered
    (1874)
139 Juewa: Chinese-based name, meaning 'Star of Fortune'
    (discovered in China)
140 Siwa: Hindu destroyer god (usually spelt Siva or Shiva in
    literary texts)
141 Lumen: Latin for 'light', name is title of book by French
    astronomer Flammarion
142 Polana: Latin for 'of Pola', town (now Pula, Yugoslavia)
    where discovered (1874)
143 Adria: Latin name of Adriatic Sea, on which Pola (see 142)
    is situated
144 Vibilia: said to be name of Roman goddess of journeyings
    (Latin *via*, 'way')
145 Adeona: said to be Roman goddess of homecomings
    (compare 144)
146 Lucina: Roman goddess of childbirth, daughter of Jupiter
    and Juno (compare 147)
147 Protogenia: daughter of Deucalion and Pyrrha; her name
    means 'first-born'
148 Gallia: Latin name of Gaul (modern France)
149 Medusa: mortal woman transformed by Athena into one of
    three Gorgons (see 681)
150 Nuwa: character in Chinese mythology; discoverer had just
    been to China
151 Abundantia: Latin for 'abundance', 'affluence'; reason for
    name not known
152 Atala: apparently for heroine of Chateaubriand's tale (1801)
    of same name
153 Hilda: one of twelve Valkyries in Norse mythology
154 Bertha: first name (Berthe) of sister of discoverer,
    Flammarion (see 169)
155 Scylla: sea nymph transformed into a sea monster (see 388
    for Charybdis)
156 Xanthippe: wife of Socrates, archetypal shrew
157 Dejanira: sister of Meleager and wife of Hercules
158 Coronis: maiden heroine of Thessaly, loved by Apollo

159 Aemilia: name of Roman noble family, or Latin version of 'Emily'

160 Una: either symbolically, as Latin for 'one', or for person of this name

161 Athor: Egyptian goddess identified with Isis (compare 42)

162 Laurentia: named for French astronomer Laurent, discoverer of Nemausa (51)

163 Erigone: daughter of Icarius (or of Aegisthus and Clytemnestra)

164 Eva: Adam's wife (biblical name)

165 Loreley: in German legend, siren (Lorelei) luring sailors to death on Rhine

166 Rhodope: girl turned into spring by Artemis when she lost virginity

167 Urda: one of three Norns in Norse mythology

168 Sibylla: one of sibyls (oracles or prophetesses)

169 Zelia: first name (Zélie) of niece of Flammarion, and daughter of his sister Berthe (see 154)

170 Maria: first name of sister of Italian astronomer Abetti, its discoverer

171 Ophelia: daughter of Polonius in Shakespeare's *Hamlet*

172 Baucis: peasant woman who with husband Philemon entertained gods

173 Ino: daughter of Cadmus whose name changed to Leucothea (see 35)

174 Phaedra: daughter of Minos and Pasiphaë, and wife of Theseus

175 Andromache: wife of Hector

176 Iduna: in Norse mythology, keeper of apples that preserved youth of gods

177 Irma: probably not from mythology, but for woman/girl of the name

178 Belisana: perhaps for Belisama, name of Minerva among Gauls (see 93)

179 Clytemnestra: wife of Agamemnon, whom she killed after Trojan War (see 911)

180 Garumna: Latin name of (river) Garonne, France, on which was discovered

181 Eucharis: nymph of goddess Calypso loved by Telemache

182 Elsa: perhaps for bride of Lohengrin in Wagner's opera of this name

183 Istria: ancient name of peninsula in Adriatic (now in Yugoslavia)

184 Deiopeia: fairest of fourteen nymphs who attended Juno

185 Eunice: name (Greek for 'good victory') is probably more symbolic than mythological, and was given to mark end of Russo-Turkish War (1878)

186 Celuta: unidentified

187 Lamberta: probably for German physicist Johann Lambert (died 1777); he has crater on Moon named after him (see Appendix I), also 'lambert', unit of brightness

188 Menippe: one of Nereids (or a daughter of Orion)

189 Phthia: Nymph seduced by Zeus (Jupiter) disguised as a pigeon

190 Ismena: daughter of Oedipus and Jocasta

191 Kolga: in Norse mythology, goddess personifying a wave

192 Nausikaa: daughter of Alcinous who helped shipwrecked Odysseus (see 197)

193 Ambrosia: the food of the gods that bestowed immortality

194 Procne: princess of Athens who was changed into a swallow on her death (see 196)

195 Eurykleia: mother of Oedipus (or nurse of Odysseus)

196 Philomena: sister of Procne (see 194), changed into a nightingale

197 Arete: wife of Alcinous and mother of Nausikaa (see 192)

198 Ampella: youth (Ampelus) loved by Dionysus (Bacchus)

199 Byblis: daughter of Miletus and granddaughter of Minos

200 Dynamene: one of Nereids

201 Penelope: wife of Odysseus, remaining true to him during his long absence

202 Chryseis: daughter of Chryses, priest of Apollo

203 Pompeia: for Pompeii, ancient city buried by eruption of Vesuvius in 79 AD; planet was discovered on same day (24 August) that town was destroyed

204 Callisto: nymph loved by Zeus (Jupiter) and turned into a bear by Hera

205 Martha: discoverer was given kindly welcome in Berlin (biblical Martha ministered to Jesus and name is that of patron saint of housewives)

206 Hersilia: Roman heroine of time of Romulus, and one of Sabine women

207 Hedda: identity unknown; too early for heroine of Ibsen's *Hedda Gabler*

208 Lacrimosa: Latin for 'weeping'; perhaps literary reference to French *comédie larmoyante* ('sentimental comedy' of eighteenth century), as named in Paris

209 Dido: Queen of Carthage who killed herself when abandoned by Aeneas

210 Isabella: identity unknown; perhaps one of historic queens of the name?

211 Isolda: princess in love with (or wife of) Tristan, also known as Yseult

212 Medea: princess who helped Jason get Golden Fleece from her father

213 Lilaea: perhaps ancient town of Achaea so named

214 Aschera: alternative name for Phoenician goddess of fertility, Astarte

215 Oenone: nymph of Mount Ida, abandoned by her lover Paris for Helen

216 Cleopatra: beautiful Egyptian queen who was mistress of Caesar

217 Eudora: son (Eudorus) of Hermes and Polymela

218 Bianca: identity unknown; perhaps woman/girl known to namer(s)

219 Thusnelda: wife of Arminius, German national hero of first century AD

220 Stephania: named for Princess Stephanie of Belgium, who in year of discovery (1881) married Archduke Rudolf, Crown Prince of Austria

221 Eos: Greek goddess of the dawn (Roman Aurora; compare 94)

222 Lucia: identity unknown, but namer was Count Hans Wilczek, sponsor of Austro-Hungarian Arctic expedition of 1872–4 (see 229)

223 Rosa: identity unknown; perhaps member of namer's family

224 Oceana: Greek god (Oceanus) who personified water that surrounded the world

225 Henrietta: first name (Henriette) of wife of French astronomer Pierre Jansen

226 Weringia: Latin name of Währing, district of Vienna where University Observatory is located; planet was named by Observatory's builder (Oberwimmer)

227 Philosophia: personification of 'love of wisdom' (philosophy; compare 275)

228 Agathe: identity unknown; namer was astronomer Theodor von Oppolzer of Vienna

229 Adelinda: first name of wife of Hans Wilczek (see 222)

230 Athamantis: daughter of Athamas, King of Thebes

231 Windobona: Latin name of Vienna, where discovered (1883)

232 Russia: named for country of discoverer, Baron von Engelhardt, who donated instruments from his observatory in Dresden to found present observatory of Kazan University, still named after him

233 Asterope: one of seven *Pleiades*

234 Barbara: identity unknown, but namer was C.H.F. Peters, who discovered forty-eight asteroids

235 Carolina: named by its discoverer Palisa, after he had observed a solar eclipse (in 1883) on Caroline Island, atoll of Line Islands, in Pacific

236 Honoria: personification of honour

237 Coelestina: origin unknown, but Celestine means 'heavenly'

238 Hypatia: perhaps for fifth-century philosopher, famous for her beauty

239 Adrastea: one of Oceanids who nursed Zeus (Jupiter)

240 Vanadis: goddess in Scandinavian mythology

241 Germania: Latin name of Germany, where discovered

242 Kriemhild: sister of Gunther wooed by Siegfried in German *Nibelungenlied*

243 Ida: mountain in Crete where Zeus (Jupiter) spent his childhood, as well as name of daughter of Melisseus (and of Corybas)

244 Sita: wife of Rama in Hindu epic poem *Ramayana*

245 Vera: identity unknown; could be Latin for 'true' as well as woman's name

246 Asporina: mother of the gods (Asporena) worshipped in Asia Minor

247 Eucrate: one of Nereids

248 Lameia: daughter (Lamia) of Poseidon (Neptune)

249 Ilse: unidentified; perhaps name of some character in German legend

250 *Bettina*

251 Sophia: unidentified; perhaps personification of cleverness (from Greek)

252 Clementina: unidentified; perhaps personification of mercy (Latin *clementia*)

253 Mathilde: unidentified; name could belong to German
   legend
254 Augusta: first name of widow of Austrian astronomer Karl
   Ludwig von Littrow
255 Oppawia: Latin name of Opava (Troppau) where discoverer
   (Palisa) was born
256 Walpurga: Almost certainly name of eighth-century abbess
   Walpurga, who gave her own name to German
   Walpurgisnacht, 'witches' sabbath' held every May Day
257 Silesia: Prussian province that was native region of
   discoverer
258 Tyche: Greek goddess of fortune (Roman Fortuna)
259 Aletheia: personification of truth, sincerity (from Greek)
260 Huberta: for St Hubert, patron saint of hunters
261 Prymno: one of Oceanids
262 Valda: identity unknown; name was proposed by Baroness
   Bettina von Rothschild (see *Bettina*)
263 Dresda: Latin name of Dresden, German city where named
   or discovered
264 Libussa: legendary princess who founded Prague (and who
   was heroine of one of Smetana's operas, *Libuse*)
265 Anna: identity unknown
266 Aline: identity unknown; name is German diminutive of
   Alexandra
267 Tirza: unidentified, although there is a biblical Tirzah (in I
   and II Kings)
268 Adorea: personification of glory; Latin *adorea* was word for
   gift of corn given soldiers after victory
269 Justitia: personification of justice
270 Anahita: goddess of fertility in Persian mythology
271 Penthesilea: Queen of the Amazons, loved (and slain) by
   Achilles
272 Antonia: identity unknown
273 Atropos: one of three Fates, who severed thread of life (see
   97 and 120)
274 Philagoria: name of recreation club in Vienna (Greek for
   'fond of assembly')
275 Sapientia: personification of wisdom (Latin; compare 227)
276 Adelheid: unidentified; name is German
277 Elvira: unidentified; Elvire is found prominently in French
   literature
278 Paulina: unidentified

279 Thule: name of 'northernmost land' in ancient geography; originally remotest known asteroid
280 Philia: either personification of love (from Greek) or some Greek nymph
281 Lucretia: for Caroline Lucretia Herschel, sister of William Herschel
282 Clorinde: identity uncertain, but in Tasso's *Jerusalem Delivered* Clorinda was the name of the daughter of the King of Ethiopia who was loved by Tancred
283 Emma: identity unknown; conceivably the heroine of Jane Austen's novel
284 Amalia: identity unknown; name is German or Italian form of 'Emily'
285 Regina: identity surprisingly unknown; perhaps for some queen (from Latin)?
286 Iclea: said to be named for the heroine of Flammarion's romance *Uranie*
287 Nephthys: Egyptian goddess (in Greek guise), sister of Isis
288 Glauke: a Nereid (or an Arcadian nymph, or the daughter of Creon of Thebes); name was selected from list by son of German astronomer Robert Luther
289 Nenetta: unidentified, but doubtless the name of a woman/girl
290 Bruna: unidentified; German name (short form of Brunhilde)
291 Alice: not known; late enough to be Wonderland Alice, but not likely to be as named by Société Astronomique de France (as was 292, both in 1889)
292 Ludovica: not known; perhaps royal name, as 291 may have been
293 Brasilia: Latin name of Brazil; name may have been prompted by abdication (1889) of Dom Pedro II, Emperor of Brazil, who was sent to Europe
294 Felicia: perhaps woman's name, but also suggests 'lucky' or 'happy'
295 Theresia: apparently for Maria Theresa (German: Maria Theresia), Queen of Hungary and Bohemia, although she had died some time earlier (in 1780)
296 Phaëtusa: one of daughters of Helios, turned into a poplar tree (compare 393)
297 Caecilia: unidentified, although there were various Roman women of the name

298 Baptistina: unknown; perhaps connected with a newly christened child?

299 Thora: feminine form of Thor, Norse god of thunder

300 Geraldina: unidentified

301 Bavaria: named by American astronomer Benjamin A. Gould to commemorate astronomical meeting just held in Munich (then capital of Bavaria)

302 Clarissa: unknown; a literary name, but not necessarily a literary link here

303 Josephina: apparently named for a person close to the Italian namer

304 Olga: said to be name of niece of German astronomer Argelander, although discovered by Palisa

305 Gordonia: perhaps tribute to James Gordon Bennett, jun., editor of *New York Herald*, who had partly financed Flammarion's observatory at Juvisy

306 Unitas: Latin for 'unity'; name commemorated unity of Italy (from 1861)

307 Nike: Greek goddess of victory; perhaps also intended to suggest French town of Nice, where was discovered

308 Polyxo: widow of Tlepolemus, King of Rhodes, who (in one account) had Helen (of Troy) hanged on tree to revenge death of her husband in Trojan War

309 Fraternitas: Latin for 'brotherhood', of which name is personification

310 Margarita: identity unknown

311 Claudia: identity unknown

312 Pierretta: identity unknown; these three names were probably individual women

313 Chaldaea: ancient region of Babylonia; Babylonians were famous astronomers (and astrologers); name was suggested by Catherine Bruce (see *Brucia*)

314 Rosalia: identity unknown

315 Constantia: either personification of constancy, or name of woman

316 Goberta: identity unknown; unusual name is perhaps based on French surname Gobert

317 Roxane: identity unknown, although Roxane was jealous wife of Alexander the Great

318 Magdalena: unknown; either personal name or perhaps tribute to biblical Mary Magdalene

319 Leona: identity unknown

320 Katharina: identity unknown, but probably relative of
    discoverer

321 Florentina: first name of daughter of discoverer

322 Phaeo: identity unknown; name looks mythological, but is
    not traceable; it suggests 'shining' in Greek

323 *Brucia*

324 Bamberga: named for astronomical meeting in Bamberg,
    Bavaria

325 Heidelberga: named for Heidelberg, German city where
    discovered (1892)

326 Tamara: Queen of Georgia in twelfth century, legendary
    romantic figure; said by some to have been named by
    Russian Grand Duke Georgy Alexandrovich

327 Columbia: tribute to Christopher Columbus; planet was
    named in year (1892) of 400th anniversary of his
    discovery of New World (compare 334)

328 Gudrun: in Scandinavian mythology, daughter of King of
    Niblungs who married Sigurd

329 Suea: name suggests Latin for 'Sweden' (Suecia); not likely
    to be for 'Swabia' since this was allocated to 417

330 Adalberta: named for father-in-law of 'discoverer' Max
    Wolf, Adalbert Merx, but planet subsequently proved
    never to have existed, and name will be reallocated in due
    course

331 Etheridgea: not clear which Etheridge this commemorates;
    perhaps either Robert Etheridge (1819–1903), English
    palaeontologist, or his namesake (1846–1920), English-
    Australian scientist; or perhaps someone else

332 Siri: origin unknown; as German name is short form of
    Sigrid (Siegrid)

333 Badenia: for Baden, (former) state where discovered (in
    Heidelberg)

334 Chicago: named at an astronomical congress in this city,
    where also Catherine Bruce (see *Brucia*) was born; name
    was to have been Columbia, for World Columbian
    Exposition of 1893, but this name already given (see 327)

335 Roberta: named for friend and benefactor, Robert von der
    Osten-Sacken, of discoverer, who was still high school
    student at time

336 Lacadiera: for French village (apparently Lacadée) where
    discoverer owned property

337 Devosa: unidentified; perhaps from Flemish surname Devos

338 Budrosa: unidentified; perhaps Latin-style version of some place-name?

339 Dorothea: for Mrs Dorothy K. Roberts, wife of Isaac Roberts, Welsh-born amateur astronomer (1829–1904); an American, she was first woman to gain degree of Doctor of Mathematical Sciences at Sorbonne, Paris

340 Eduarda: for German amateur astronomer, Eduard von Lade (1817–1904)

341 California: so named by Max Wolf, who had visited this American state in 1893 (and who seemed to have deliberately started a new name strain)

342 Endymion: beautiful youth visited nightly by moon goddess Selene

343 Ostara: Germanic goddess of the sun

344 Desiderata: named for wife of King Bernadotte, founder of present royal Swedish line (from 1818)

345 Tercidina: unidentified; name seems too contrived to be personal name

346 Hermentaria: apparently for French village of Herment (Puy-de-Dôme)

347 Pariana: unidentified, although this is name of Vedic goddess of fertility

348 May: unidentified; perhaps for an American (or English) woman?

349 Dembowska: for Italian astronomer, Baron Ercole Dembowski

350 Ornamenta: said to be for a Dutch sailor named Horneman, whose daughter was member of Société Astronomique de France

351 Yrsa: unidentified; unusual enough to be a particular personal name

352 Gisela: first name of wife of German astronomer Max Wolf

353 Ruperto-Carola: Latin version of name of Heidelberg University (Ruprecht-Karl-Universität), founded in fourteenth century by Rupert I, Count Palatine of the Rhine

354 Eleonora: identity unknown

355 Gabriella: identity unknown (but compare *Gabi*)

356 Liguria: named for region of north-west Italy

357 Ninina: origin unknown

358 Apollonia: perhaps meant to suggest Apollo?

359 Georgia: not clear whether based on place or personal name

360 Carlova: perhaps meant to suggest Charles; discoverer was Charlois (1893)
361 Bononia: Latin name of Bologna, Italy
362 Havnia: Latin name of Copenhagen, Denmark
363 Padua: named for Italian city
364 Isara: Latin name of French river Isère, tributary of Rhône
365 Cordova: named for Spanish city (Córdoba)
366 Vincentina: unidentified; perhaps from personal name
367 Amicitia: personification of friendship (from Latin)
368 Haidea: perhaps for the Haidee in Byron's *Don Juan*
369 Aëria: one of names of Juno, regarded as 'queen of the air'
370 Modestia: personification of modesty (Latin)
371 Bohemia: named for (former) central European kingdom
372 Palma: probably for 'palm' in symbolic or literal (botanical) sense
373 Melusina: probably named for famous fairy of French folklore
374 Burgundia: Latin name of French historic province of Burgundy
375 Ursula: probably particular personal name
376 Geometria: personification (or symbolic name) of geometry
377 Campania: named for region of south-west Italy
378 Holmia: Latin name of Stockholm, Sweden
379 Huenna: unidentified, although Huena was Latin name of Swedish island of Ven (Hven), home of famous astronomer Tycho Brahe
380 Fiducia: personification of confidence, trust (Latin)
381 Myrrha: mother of Adonis, who was changed into a myrtle by Aphrodite
382 Dodona: ancient town of Thessaly where there was an oracle to Jupiter
383 Janina: unidentified; based on personal name?
384 Burdigala: Latin name of Bordeaux, France, where planet was discovered (1894)
385 Ilmatar: in Finnish mythology, a nymph of the air
386 Siegena: Latin-style name of Siegen, German city east of Cologne
387 Aquitania: for Aquitaine, historic province of south-west France; planet was discovered in Bordeaux (see 384), which is in region
388 Charybdis: ship-devouring monster in Greek mythology (for Scylla see 155)

389 Industria: personification of industry, ability

390 Alma: unidentified; hardly likely to be for Battle of Alma
in Crimean War, as this was in 1854, unless was intended
to mark its fortieth anniversary; perhaps simply Latin for
'bountiful' or 'life-giving' (as used of deities)?

391 Ingeborg: perhaps for thirteenth-century Danish princess
who became Queen of France

392 Wilhelmina: perhaps for recently enthroned Queen of the
Netherlands

393 Lampetia: one of daughters of Helios changed into a poplar
tree (compare 296)

394 Arduina: Gaulish version of name of goddess of hunting,
Diana (compare 78)

395 Delia: an alternative name for Diana (see 394), from island
of Delos

396 Aeolia: ancient country of Asia Minor, home of wind god,
Aeolus

397 Vienna: not Austrian capital but French city of Vienne;
name given (rather misleadingly) by French discoverer
Charlois to commemorate native land

398 Admete: son (Admetus) of Pheres, King of Thessaly, who
married Alceste

399 Persephone: daughter of Zeus (Jupiter) and Demeter
(*Ceres*) who was abducted by Hades and taken to be
queen of underworld (Roman Proserpina; see 26)

400 Ducrosa: presumably from French surname Ducros

401 Ottilia: unidentified; name seems contrived

402 Chloë: alternative name of Demeter (Roman *Ceres*)

403 Cyane: Sicilian nymph changed into a fountain by Pluto
(Hades)

404 Arsinoë: daughter of Leucippus (or of Phegeus, King of
Psophis)

405 Thia: mother of Sun, Moon and Aurora (according to some
accounts)

406 Erna: not a mythological name, but presumably of a
(probably German) woman

407 Arachne: mythological woman adept at weaving and
changed into a spider

408 Fama: goddess who was daughter of the Sun and the Earth

409 Aspasia: Greek beauty of fifth century BC who lived with
Pericles

410 Chloris: daughter of Amphion and Niobe; her name means 'pale'

411 Xanthe: river of Troy or its god (Xanthus); his name means 'yellow'

412 Elisabetha: unidentified

413 Edburga: unidentified, although Eadburga was ninth-century Queen of West Saxons

414 Liriope: one of Oceanids, mother of Narcissus

415 Palatia: apparently for Palatine, one of Seven Hills of Rome (compare 416)

416 Vaticana: perhaps not so much Vatican, papal palace in Rome, as hill on which built there (compare 415)

417 Suevia: Latin name of Swabia (German Schwaben), former German duchy

418 Alemannia: Latin name of Germany (compare French Allemagne)

419 Aurelia: probably from Aurelius, Roman clan name

420 Bertholda: perhaps for German monk, Meister Berthold, or a royal (ducal) name; probably latter in view of 421

421 Zähringia: for Zähringen, archducal manor in south-west Germany, title of many dukes named Berthold (compare 420), with head of house eleventh-century Berthold I.

422 Berolina: feminine form of Berolinum, Latin name of Berlin, where discovered

423 Diotima: Greek priestess of fifth century BC (said to have taught Socrates)

424 Gratia: personification of charm or grace, or reference to Three Graces

425 Cornelia: unidentified, but perhaps for one of famous Roman women of the name, such as mother of the Gracchi (second century BC) or wife of Julius Caesar

426 Hippo: probably for ancient port of Roman Africa, home of St Augustine

427 Galene: for Galen, famous Greek physician of second century AD

428 Monachia: feminine form of Monachium, Latin name of Munich, Germany, where planet was discovered

429 Lotis: nymph, daughter of Neptune; changed into lotus to escape Priapus

430 Hybris: hardly likely to be personification of violence (Greek); perhaps some ancient place-name

431 Nephele: wife of mythical King Athamas of Boeotia
   (compare 435)
432 Pythia: Apollo's prophetess, an oracular snake (python)
433 *Eros*
434 Hungaria: Latin name of Hungary; named to mark
   astronomical congress at Budapest that year (1898)
435 Ella: daughter of King Athamas (compare 431)
436 Patricia: probably reference to Roman patrician families
   (aristocrats)
437 Rhodia: probably reference to ancient Greek island or city
   of Rhodes
438 Zeuxo: one of Oceanids (daughters of Oceanus)
439 Ohio: named for American state where discovered (compare
   440)
440 Theodora: for the daughter of Julius F. Stone, of Ohio State
   University, who was now (1898) married to Professor
   Charles Sutton there (see also 445)
441 Bathilde: perhaps for wife of Clovis II, seventh-century King
   of France
442 Eichsfeldia: for Eichsfeld, district of Thuringia (now East
   Germany)
443 Photographica: named to mark new method of discovering
   minor planets, although this was not the first to be so
   photographed (which was *Brucia*)
444 Gyptis: in legend, the wife of Protis, said to be Greek leader
   of expedition to site where French city of Marseille was
   founded (in about 600 BC)
445 Edna: first name of wife of Julius F. Stone (see 440); she
   had made several gifts to promote scientific work at Ohio
   State University
446 Aeternitas: personification of eternity
447 Valentine: unidentified, but perhaps reference to one of
   (female) characters in Meyerbeer's opera *Les
   Huguenots*, although this was written over sixty years
   earlier
448 Natalie: unidentified; name suggests connection with
   Christmas, but quite likely to be woman's name
449 Hamburga: for Hamburg, to mark annual meeting there of
   Mathematical Society (compare 454)
450 Brigitta: probably name of particular woman/girl known to
   discoverer
451 *Patientia*

452 Hamiltonia: for Mount Hamilton, California, on whose summit is Lick Observatory, where planet was discovered (1900)

453 Tea: Irish goddess, patroness of Tara

454 Mathesis: symbolic of desire of (or power of) learning (Greek word); name was given during same mathematical meeting that gave 449

455 Bruchsalia: for West German city of Bruchsal

456 Abnoba: Celtic (Gaulish) divinity similar to Diana (compare 394)

457 Alleghenia: for Allegheny Observatory of University of Pittsburgh, Pennsylvania; city is itself seat of Allegheny County (itself named after river; compare 484)

458 Hercynia: Latin name (Hercynia Silva) of forest-covered mountain region of Germany and eastern Europe

459 Signe: unidentified; perhaps some ancient place-name such as Signia?

460 Scania: Latin name of southern part of Sweden (modern Skåne)

461 Saskia: unidentified; perhaps a tribute to Rembrandt's wife?

462 Eriphyla: wife of Amphiaraüs who denounced her husband and was killed by her son Alcmeon

463 Lola: unidentified; perhaps for character in Mascagni's opera *Cavalleria Rusticana*, produced ten years earlier

464 Megaira: one of three Furies

465 Alekto: second of the Furies

466 Tisiphone: third of the Furies

467 Laura: unidentified; perhaps real woman/girl rather than mythological

468 Lina: unidentified; probably familiar name of woman/girl with longer name such as Carolina, Angelina, Adelina

469 Argentina: for the country

470 Kilia: unidentified

471 Papagena: name almost certainly that of the character in Mozart's opera *The Magic Flute* (where she is loved by Papageno) (compare 539)

472 Roma: for Italian capital, of which discoverer (Carnera, 1901) was a native (compare 477)

473 Nolli: unidentified; name suggests familiar form of 'Oliver'

474 Prudentia: personification of prudence (Latin); (compare Greek form, 134)

475 Ocllo: said to have been named by discoverer (Stewart,

1901) for first Queen of Incas, regarded as daughter of the Sun

476 Hedwig: unidentified; perhaps one of historic European women of the name

477 Italia: for native country of discoverer (Carnera, 1901) (compare 472)

478 Tergeste: Latin name of Italian city of Trieste

479 Caprera: name of island off north-east Sardinia, and home of Garibaldi, to whom perhaps was tribute

480 Hansa: for medieval guild of merchants who gave name to Hanseatic League

481 Emita: unidentified; perhaps personal name rather than place-name

482 Petrina: probably personal name, but individual unidentified

483 Seppina: based on name of Sepp, German astronomer Max Wolf's dog (!); (Sepp is familiar form of 'Joseph' in German)

484 Pittsburghia: for Pittsburgh, Pennsylvania (compare 457)

485 Genua: for the Italian city (Genoa)

486 Cremona: for the Italian city

487 Venetia: for Venice, Italy

488 Creusa: daughter of Erechtheus (or of Priam and Hecuba)

489 Comacina: name of sole island on Lake Como, Italy

490 Veritas: personification of truth (Latin)

491 Carina: unidentified; perhaps ancient place-name or mythological name

492 Gismonda: apparently for Gismonde, daughter of Tancred and lover of Guiscard in medieval tale (retold in Boccaccio's *Decameron*); name is too early to be taken from French opera *Gismonde* by Février (compare 493)

493 Griseldis: probably from patient woman (Griselda) in medieval tale retold by Boccaccio in *Decameron*, but could have been taken from Massenet's opera *Grisélidis*, first performed previous year (1901; compare 492)

494 Virtus: personification of virtue; name was proposed by Flammarion, who regretted it had not been used earlier

495 Eulalia: probably personal name of woman/girl, although unidentified

496 Gryphia: apparently for gryphon (griffin), winged monster with head of eagle and body of lion

497 Iva: unidentified; may even refer to male named Ivo (Italian or German name)

498 Tokio: for Japanese capital

499 Venusia: does not seem to be reference to Venus; perhaps for ancient town of Apulia, where Roman poet Horace was born, and so even tribute to him

500 Selinur: unidentified; name seems to be from Mediterranean rather than Scandinavia, as do 501 and 502

501 Urhixidur: unidentified (see 500)

502 Sigune: unidentified, although possibly linking with Sigunae, ancient Scythian tribe (see 500)

503 Evelyn: unidentified

504 Cora: said to have been taken by namer (Pickering) from Peruvian mythology, as was 505

505 Cava: perhaps from Peruvian mythology (see 504)

506 Marion: unidentified

507 Laodica: daughter of Priam and Hecuba (or of Agamemnon and Clytemnestra)

508 Princetonia: for Princeton, New Jersey, where Halsted Observatory is situated (as part of university; compare 534)

509 Iolanda: unidentified; name may have literary link

510 Mabella: unidentified; name could be blend of 'Mabel' and 'Bella'

511 Davida: appears to refer to King David of the Bible

512 Taurinensis: refers to Taurini, former tribe of Liguria, Italy

513 Centesima: Italian for 'hundredth'; planet was hundredth discovered by Max Wolf

514 Armida: perhaps from title of opera by either Lully or Gluck, with name originally that of heroine of Tasso's *Jerusalem Delivered*

515 Athalia: perhaps biblical character who is the heroine of Racine's play *Athalie*

516 Amherstia: named for observatory of Amherst College, Amherst, Massachusetts, founded in 1821 to train men for the ministry

517 Edith: identity unknown; perhaps member of discoverer's family

518 *Halawe*

519 Sylvania: perhaps feminine form of Sylvanus, Roman god of woods

520 Franziska: feminine form of Franziskus, name (German for 'Francis') of one of sons of Max Wolf

521 Brixia: Latin name of Brescia, town in northern Italy

522 Helga: unidentified
523 Ada: unidentified; perhaps historic name
524 Fidelio: appears to be named for Beethoven's opera, or of character in it (who is Leonora in male disguise; see 696)
525 Adelaide; unidentified; probably not queen of William IV of Britain, or Adelaide, Australia; in view of run of names here, perhaps opera character?
526 Jena: German city
527 Euryanthe: probably heroine of Weber's opera of same name
528 Rezia: probably for character in Weber's opera *Oberon*
529 Preziosa: perhaps for character in Cervantes' novel, *La Gitanilla*
530 Turandot: too early for heroine of Puccini's opera of the name, first performed in 1926, but princess's name was familiar from Gozzi's play of 1762
531 Zerlina: probably for character in Mozart's opera *Don Giovanni*
532 Herculina: unidentified; perhaps a literary or operatic character?
533 Sara: unidentified, unless biblical character of the name
534 Nassovia: Latin name of Nassau, with special reference to University of Princeton, New Jersey, whose first building was named Nassau Hall 'in memory of the glorious King William III' (compare 508)
535 Montague: identity unknown
536 Merapi: presumably for the volcano of the name in Java, Indonesia
537 Pauly: said to be for a friend of Max Wolf
538 Friederike: unidentified; name of a German woman/girl
539 Pamina: unidentified, but in view of operatic names above and below could well be the Pamina in Mozart's opera *The Magic Flute* (compare 471)
540 Rosamunde: unidentified; name is familiar in literature and music (Schubert)
541 Deborah: unidentified; perhaps biblical character, or else one of many women of name in later literature
542 Susanna: unidentified; perhaps a German woman/girl
543 Charlotte: unidentified; perhaps (like 542) a German name
544 Jetta: unidentified, although associated with a Heidelberg legend; name is short form of Henriette

545 Messalina: probably for famous wife of Roman emperor
     Claudius
546 Herodias: probably for wife of Herod Antipas
547 Praxedis: unidentified; perhaps literary character
548 Cressida: woman associated with Troilus in many literary
     accounts
549 Jessonda: unidentified; name seems to be a blend or
     concoction, e.g. of 'Jessica' and 'Rosamonda'
550 Senta: perhaps for character (sea captain's daughter) in
     Wagner's opera *The Flying Dutchman*, especially in view
     of next three names
551 Ortrud: Count of Brabant's wife in Wagner's opera
     Lohengrin
552 Sigelinde: Wotan's daughter in Wagner's opera *The Valkyrie*
553 Kundry: sorceress in Wagner's opera *Parsifal*
554 Peraga: unidentified; perhaps also an operatic character?
555 Norma: almost certainly for heroine of Bellini's opera of
     this name
556 Phyllis: either for girl in Greek mythology who married
     Acamas, or for an operatic character, or for the classical
     rustic girl in literature
557 Violetta: heroine of Verdi's opera *La Traviata*
558 Carmen: heroine of Bizet's opera of the same name
559 Nanon: perhaps similarly for an operatic or literary
     character?
560 Delila: heroine of opera by Saint-Saëns, *Samson et Dalila*
561 Ingwelde: unidentified; also operatic?
562 Salome: heroine of Richard Strauss's opera of same name
563 Suleika: perhaps for character in Goethe's *Der West-östliche
     Divan*
564 Dudu: perhaps for character in Nietzsche's *Thus Spake
     Zarathustra*
565 Marbachia: probably for Marbach, German town where
     Schiller was born
566 Stereoscopia: named by German inventor (Pulfrich) of
     'stereo-comparator', photographic device for giving
     stereoscopic image of viewed planet
567 Eleutheria: personification of liberty (and Greek goddess of
     it)
568 Cheruskia: said to be name of students' association in
     Heidelberg
569 Misa: perhaps a mythological name of some kind?

570 Kythera: island in Aegean that was centre of cult of
  Aphrodite
571 Dulcinea: heroine of Cervantes' *Don Quixote* (peasant girl
  loved by him)
572 Rebecca: unidentified; perhaps the biblical wife of Isaac.
573 Recha: unidentified, unless it is biblical name of
  I Chronicles 4:12
574 Reginhild: unidentified; more a Scandinavian name than a
  German one
575 Renate: unidentified; a German name
576 Emanuela: unidentified; a German name
577 Rhea: one of Titanesses in Greek mythology
578 Happelia: for benefactor of Königstuhl Observatory, near
  Heidelberg
579 Sidonia: unidentified; perhaps personal name rather than
  place-name
580 Selene: goddess of the moon
581 Tauntonia: for city of Taunton, Massachusetts, where
  discovered
582 Olympia: for famous city of ancient Greece, home of
  Olympic Games, or even to mark modern Olympic
  Games, which by now were under way
583 Clotilde: first name of daughter of Edmund Weiss,
  astronomer at Vienna
584 Semiramis: either legendary Greek queen, or historical
  Assyrian one
585 Bilkis: unidentified; name of legendary character?
586 Thecla: saint, said to have been first woman martyr
587 Hypsipyle: Queen of Lemnos who bore sons to Jason
588 *Achilles*
589 Croatia: historic region corresponding to modern republic
  of Yugoslavia
590 Tomyris: Queen of the Massegetae who killed invading
  Cyrus
591 Irmgard: unidentified; German name
592 Bathseba: biblical character (Bathsheba)
593 Titania: probably for queen of fairies in Shakespeare's *A
  Midsummer Night's Dream*
594 Mireille: heroine of French poet Mistral's poem *Mirèio*
  (poem is written in Provençal and heroine's name is
  French Mireille); name proposed by Flammarion
595 Polyxena: daughter of Priam and Hecuba

596 Scheila: said to be surname (Scheil) of acquaintance of
    discoverer (Kopff)
597 Bandusia: name of fountain in Apulia, Italy, near where
    classical poet Horace had a cottage
598 Octavia: probably for one of classical women of the name,
    such as daughter of Claudius and Messalina who was wife
    of Nero (and executed by him)
599 Luisa: identity unknown; German name
600 Musa: name is tribute to nine Muses, daughters of Zeus
    (Jupiter) and Mnemosyne (57): see 18, 22, 23, 27, 30,
    33, 62, 81 and 84
601 Nerthus: Scandinavian goddess of fertility
602 Marianna: unidentified
603 Timandra: daughter of Tyndareüs and Leda, and mother of
    Laodocus
604 Tekmessa: concubine of Ajax, to whom she bore a son,
    Eurysaces
605 Juvisia: for Juvisy, town in northern France, where
    Flammarion had his observatory
606 Brangane: Isolde's attendant in Wagner's opera *Tristan and
    Isolde*
607 Jenny: said to have been a bridesmaid friend of discoverer
    (Kopff), together with 608 and 616
608 Adolfine: friend of discoverer (see 607)
609 Fulvia: Roman matron, wife successively of Clodius, Curio
    and Mark Antony
610 Valeska: unidentified
611 Valeria: unidentified
612 Veronika: unidentified; a coincidence that these three names
    begin with 'V'?
613 Ginevra: unidentified; a personal name related to 'Jennifer'
    and 'Guinevere'
614 Pia: unidentified; German name (as feminime of Pius)
615 Roswitha: said to be for German poet Roswitha von
    Gandersheim, tenth-century nun who is regarded as first
    German poetess
616 Elly: named for a friend of discoverer (Kopff; see 607)
617 *Patroclus*
618 Elfriede: unidentified
619 Triberga: for German city of Triberg
620 Drakonia: for Drake University, Des Moines, Iowa, where
    orbit was first computed (compare 694)

621 Werdandi: one of three Norns (goddesses) in Norse
    mythology
622 Esther: probably for biblical heroine
623 Chimaera: mythical monster; mixture of lion, goat and snake
624 *Hector*
625 Xenia: unidentified; perhaps feminine form of Xenius, one
    of names given to Zeus (Jupiter) meaning 'hospitable'
626 Notburga: unidentified; German personal name
627 Charis: personification of grace (and name of goddess who
    was Vulcan's wife)
628 Christine: first name of wife of discoverer (Kopff)
629 Bernardina: not identified
630 Euphemia: not identified
631 Philippina: said to be fiancée of a friend of the discoverer
    (compare 634 and 668)
632 Pyrrha: first mortal-born woman, marrying her cousin
    Deucalion
633 Zelima: identity unknown
634 Ute: said to be fiancée of a friend of the discoverer (compare
    631 and 668)
635 Vundtia: tribute to German physiologist, Wilhelm Wundt
636 Erika: unidentified
637 Chrysothemis: daughter of Agamemnon and Clytemnestra
638 Moira: Greek name (Moirai) of Fates
639 Latona: Roman counterpart to Leto (see 68)
640 Brambilla: name of character in one of tales by German
    writer Hoffmann
641 Agnes: unidentified
642 Clara: unidentified
643 Scheherezade: name of (female) narrator of tales of *Arabian
    Nights*
644 Cosima: perhaps for Liszt's daughter who became Wagner's
    wife
645 Agrippina: for one of two daughters of Agrippa (first
    century AD)
646 Castalia: either name of ancient town near Phocis, or
    mythological character
647 Adelgunde: unidentified; German name
648 Pippa: unidentified; perhaps literary character, such as
    heroine of Hauptmann's play *Und Pippa tanzt*, first
    performed 1906
649 Josefa: unidentified

650 Amalasuntha: unidentified, but rather rare German name

651 Anticleia: wife of Laërtes and mother of Odysseus

652 Jubilatrix: Latin for 'woman who rejoices'; named to mark jubilee (sixtieth anniversary) of reign of Austrian emperor Franz Josef, as discovered in this year (1908)

653 Berenice: either for one of several Egyptian or Jewish princesses so named, or for literary or musical character in work about one, e.g. Racine's tragedy, Poe's tale, and so on

654 Zelinda: not identified

655 Briseïs: concubine of Achilles, and later of Agamemnon

656 Beagle: name of ship (the first in the list) in which Darwin travelled round world in 1830s

657 Gunlöd: giant's daughter in Norse mythology, seduced by Odin

658 Asteria: daughter of Titans Phoebe and Coeüs, and mother of Hecate

659 Nestor: son of Neleus and Chloris who married Eurydice; a rare male name among these almost exclusively feminine ones

660 Crescentia: unidentified; perhaps simply personification of increase (Latin)

661 Cloelia: Roman girl who escaped from King Porsenna to swim over Tiber

662 Newtonia: for Isaac Newton, famous English philosopher and mathematician, who among other things constructed reflecting telescope

663 Gerlinde: unidentified; German name

664 Judith: unidentified; perhaps literary name, as that of heroine of Hebbel's tragedy of same name (or even for biblical character, on which it was based)

665 Sabine: group of tribes living near Rome, whose women were abducted by the Romans ('Rape of the Sabines')

666 Desdemona: innocent heroine of Shakespeare's *Othello*

667 Denise: unidentified; perhaps French name rather than German

668 Dora: said to be name of woman loved by a friend of discoverer (Kopff) (compare 631 and 634)

669 Cypria: one of names of Aphrodite, who was born from sea near Cyprus

670 Ottegebe: name of character in Hauptmann's *Der Arme Heinrich*

671 Carnegia: in honour of Carnegie Institution of Washington, founded (1902) by industrial benefactor Andrew Carnegie

672 Astarte: the Middle Eastern name of Aphrodite, known to the Babylonians as Ishtar, and regarded as the daughter of the Sun and Moon

673 Edda: the name of one or other of a collection of mythological Scandinavian writings (especially poems), dating from medieval times

674 Rachele: unidentified; probably not the biblical Rachel

675 Ludmilla: unidentified; probably the name of a German woman/girl, despite its Russian associations

676 Melitta: a form of the biblical name Melissa, apparently selected to blend with the name of the discoverer, who was P.J. Melotte

677 Aaltje: unidentified; name seems Scandinavian

678 Fredegundis: name of heroine of opera *Frédégonde* by French composer Ernest Guirand, first staged (posthumously) in 1895

679 Pax: personification of peace (Latin; compare 58)

680 Genoveva: probably for Genevieve (earlier Genovefa) of Brabant, heroine of old German folk tale

681 Gorgo: name representing one of or all three monstrous sisters in Greek mythology (Gorgons), best known of whom was Medusa (see 149)

682 Hagar: probably biblical character (Sarah's Egyptian maid)

683 Lanzia: not identified; perhaps Italian name, rather than German

684 Hildburg: not identified; is recognised German name

685 Hermia: perhaps for daughter of Egeus of Athens, who escapes to woods with Lysander in Shakespeare's *A Midsummer Night's Dream*

686 Gersuind: unidentified; perhaps a literary character?

687 Tinette: unidentified

688 Melanie: unidentified

689 Zita: unidentified; all these three names could be German

690 Bratislavia: for Bratislava, city in Czechoslovakia

691 Lehigh: for American river in Pennsylvania

692 Hippodamia: one of many daughters in Greek mythology

693 Zerbinetta: coquettish character in Richard Strauss's opera *Ariadne auf Naxos*

694 Ekard: word play begins here, as name is reversal of 'Drake'

(see 620); name was given by American astronomer Seth B. Nicholson and his wife to mark computation (1908) of planet's orbit by students of Drake University

695 Bella: unidentified; perhaps literary or operatic character?

696 Leonora: very likely for one of popular operatic characters of the name, such as heroine of Verdi's *Il Trovatore*, or Florestan's wife disguised as male in Beethoven's *Fidelio* (see 524), or the favourite of the title of Donizetti's *La Favorite*, and so on

697 Galilea: for Galileo, the great Italian astronomer

698 Ernestina: feminine form of Ernest, name of son of discoverer Max Wolf

699 Hela: Queen of the Dead and daughter of Loki in Scandinavian mythology (see 949)

700 Auravictrix: Latin name means literally 'conqueress of the breeze'; perhaps this was descriptive name for some Roman goddess, or was devised as alternative to Aurora (see 94)

701 Oriola: origin unknown, but in view of 702 could be for bird oriole (see 706)

702 Alauda: Latin name of lark; both it and oriole (see 701) are songbirds

703 Noëmi: this is variant of biblical Naomi, mother-in-law of Ruth

704 Interamnia: from Latin name, Interamnium, of Italian town of either Teramo or Terni

705 Erminia: Syrian girl whose life was spared by Tancred in Tasso's *Jerusalem Delivered*

706 Hirundo: Latin name of the swallow (for other bird names see 701, 702, 709, 713 and 714)

707 Steïna: for name of astronomer (Stein) working in Riga, Latvia

708 Raphaela: perhaps for archangel Raphael, but could be name of woman/girl

709 Fringilla: Latin name of chaffinch (or finch in general; see 706)

710 Gertrud: perhaps for queen and mother of Hamlet in Shakespeare's play

711 Marmulla: not identified; perhaps literary or operatic name?

712 Boliviana: for Simón Bolívar, liberator who gave his name to Bolivia; name was proposed by Flammarion (1911)

713 Luscinia: Latin name of nightingale (see 706)

714 Ulula: Latin name of screech owl (see 706)

715 Transvaalia: for Transvaal, South Africa, where was first planet to be discovered there (1911)

716 Berkeley: for Berkeley, California, seat of University of California

717 Wisibada: Latin name of German city of Wiesbaden

718 Erida: apparently named for a sister of American astronomer Armin O. Leuschner, director of Students' Observatory of University of California (now Leuschner Observatory)

719 Albert: for first name of Albert Rothschild, Viennese financier and benefactor, who died this year (1911); planet has since been lost (compare 977)

720 Bohlinia: for Swedish astronomer Karl Bohlin, director of Stockholm Observatory

721 Tabora: named for ship (of East Africa Line) on which astronomers had come to Hamburg for meeting of German Astronomische Gesellschaft (Astronomical Society); this name appears to link up with 722, 723 and 724

722 Frieda: from German word for 'peace', *Friede* (see 721)

723 Hammonia: Latin name of Hamburg (see 721)

724 Hapag: from abbreviation that was commercial name of shipping line, Hamburg-Amerika-Linie (from German: *H*amburg-*A*merikanische *P*acketfahrt-*A*ktien-*G*esellschaft) (see 721); planet has since been lost

725 Amanda: first name of Amanda Schorr, wife of the namer

726 Joëlla: feminine form of first name of the discoverer, Revd Joël Metcalf (see 792)

727 Nipponia: for Japan, one of whose Japanese names is Nippon

728 Leonisis: unexplained

729 Watsonia: for surname of American astronomer James Craig Watson, who discovered twenty-two minor planets

730 Athanasia: possibly for St Athanasius, Greek father of the church, or else for a woman named Athanasia

731 Sorga: daughter of Oeneus, king of Calydon

732 Tjilaki: unidentified; perhaps a mythological name?

733 Mocia: for Max Wolf's youngest son Werner, nicknamed 'Mok'

734 Benda: for Czech composer of eighteenth century, Georg Benda, or possibly for nineteenth-century Czech

composer Karel Bendl; either way, planet-namer was
Palisa, born in Troppau (now Czech city of Opava)

735 Marghanna: combined name, Margaret and Anna, of
mother and wife of discoverer (compare 746)

736 Harvard: for famous American university at Cambridge,
Massachusetts, and more especially for observatory there
(compare 737, 740 and 878)

737 Arequipa: name of city in Peru where branch of Harvard
Observatory was located (compare 736)

738 Alagasta: unidentified; perhaps a place-name?

739 Mandeville: town in Jamaica, where American astronomer
E.C. Pickering made observations

740 Cantabia: not identified, but name suggests 'Cambridge',
and this must be Cambridge, Massachusetts, especially
as it has famous observatory (see 736 and 767)

741 Botolphia: for Boston, Massachusetts, which was named
after Boston, England (which was named after St
Botolph); Harvard University and other famous
observatories are not far away (compare 737, 740 and
758)

742 Edisona: for famous American physicist and inventor,
Thomas A. Edison

743 Eugenisis: unidentified

744 Aguntina: unidentified

745 Mauritia: said to be named after 'Mauritius Blue' postage
stamp

746 Marlu: unidentified; perhaps combined name (e.g.
Margarete and Luise), like 735?

747 Winchester: for city of Winchester, Massachusetts, where
discovered by American astronomer Metcalf

748 Simeïsa: for Russian town of Simeiz, in Crimea, where
discovered by Russian astronomer Grigory Neujmin (see
749, 751, 753 and 779; compare 951)

749 Malzovia: for Russian amateur astronomer N.S. Maltsov,
who presented his private observatory at Simeiz (see
748) to the Pulkovo Observatory in 1908 (where it was
renamed as the Crimean Astrophysical Observatory; see
762 and 814)

750 Oskar: unidentified; in view of names here, perhaps a
Russian link?

751 Faïna: first name of (first) wife of Russian astronomer G.N.
Neujmin (see 748 and 753)

752 Sulamitis: for heroine of story ('Sulamif' ', 1908) by Russian writer Alexander Kuprin

753 Tiflis: Russian city (now Tbilisi), capital of Georgia, where Russian astronomer Neujmin was born (see 748)

754 Malabar: for Indian district of the name

755 Quintilla: unidentified, although there was a Roman courtesan of the name

756 Lilliana: unidentified; perhaps there was some literary Lillian?

757 Portlandia: for American city of Portland, Maine

758 Mancunia: Latin name of Manchester, England, but here probably for an American town or city of the name (like 741); perhaps Manchester, Massachusetts, near Boston, or Manchester, New Hampshire?

759 Vinifera: unexplained; there is a type of European grape of the name, but this seems unlikely source; perhaps some private reference?

760 Massinga: for surname (Massinger) of assistant astronomer at Königstuhl Observatory, near Heidelberg

761 Brendelia: for German astronomer Otto Brendel, director of University Observatory at Frankfurt

762 Pulcova: for Russian Pulkovo Observatory, near Leningrad, where founded in 1839, but now with department in Crimea (see 748, 749, 768, 786, 848 and 857)

763 Cupido: for Cupid, Roman god of love

764 Gedania: Latin name of Danzig (now Gdansk. Poland), an important cultural centre

765 Mattiaca: Latin name of German city of Marburg

766 Moguntia: Latin name of German city of Mainz

767 Bondia: tribute to American astronomers William C. Bond and his son George P. Bond, who were directors of Cambridge Observatory, Massachusetts (see 740)

768 Struveana: for one or more of famous astronomer family Struve, two of whom were directors of Pulkovo Observatory (see 762)

769 Tatiana: not identified, but in view of Russian names here almost certainly for heroine of Pushkin's *Eugene Onegin*

70 Bali: probably for Indian god Bali rather than Indian town or Indonesian island of the name

771 Libera: Roman fertility goddess identified with Greek Persephone

772 Tanete: unidentified

773 Irmintraud: unidentified; perhaps this name and 772 are literary?

774 Armor: possibly for Armorica, ancient name of Brittany

775 Lumière: tribute to French Lumière brothers, chemists and cinematograph pioneers

776 Berberica: tribute to German astronomer Berberich

777 Gutemberga: for Johann Gutenberg, fifteenth-century German inventor of printing

778 Theobalda: for Theobald Weyres, assistant astronomer at Königstuhl Observatory, near Heidelberg

779 Nina: first name of sister of Russian astronomer G.N. Neujmin (see 748 and 751)

780 Armenia: for ancient Asian kingdom, now republic of Soviet Union (see 791)

781 Kartvelia: native name of Russian republic of Georgia

782 Montefiore: for Arthur Montefiore, secretary of Arctic expedition of 1890s led by British explorer F.G. Jackson

783 Nora: unidentified

784 Pickeringia: tribute to American astronomer brothers Edward C. Pickering and William H. Pickering

785 Zwetana: unidentified; perhaps Russian name Tsvetana (compare 825) and member of family of one of astronomers at Simeiz (see 748), or literary character

786 Bredichina: tribute to Russian director of Pulkovo Observatory (see 762), F.A. Bredikhin

787 Moskva: Russian name of Moscow, as general tribute to Russian astronomers

788 Hohensteina: for Hohenstein, German city (now in East Germany)

789 Lena: famous Russian river, or possibly Russian woman's name?

790 Pretoria: capital of Transvaal, South Africa, and of South Africa itself

791 Ani: name of ancient political and cultural centre of Armenia (see 780)

792 Metcalfia: for Joël H. Metcalf, American astronomer, who discovered several minor planets (see 726)

793 Arizona: American state, where famous Lowell Observatory is situated

794 Irenaea: unidentified

795 Fini: unidentified

796 Sarita: unidentified, although there is a place of the name
   in Texas (but it was based on a woman's first name)
797 Montana: apparently not American state, but Latin-style
   name of German town of Bergedorf (Latin *mons* and
   German *Berg* both mean 'mountain'), where planet was
   discovered at observatory
798 Ruth: presumably for biblical heroine (after whom Old
   Testament book is named)
799 Gudula: apparently named arbitrarily by German
   astronomer Karl W. Reinmuth, assistant to Max Wolf;
   for further arbitrary names devised by him, see examples
   between 913 and 929
800 Kressmannia: for surname (Kressmann) of donor of
   refractor telescope to Königstuhl Observatory, near
   Heidelberg
801 Helvetia: Latin name of Switzerland
802 Epyaxa: name of an early Queen of Cilicia (Asia Minor)
803 Picka: for the Czech doctor, Friedl Pick (1867–1921)
804 Hispania: Latin name of Spain; planet was first discovered
   there
805 Hormuthia: for maiden name (Hormuth) of wife of
   astronomer Professor Wolff
806 Gyldenia: for Hugo Gylden, Swedish astronomer and
   director of Stockholm Observatory
807 Ceraskia: for Vitold Karlovich Tserasky, director of Moscow
   Observatory from 1890 to 1916
808 Merxia: for maiden name (Merx) of wife of German
   astronomer Max Wolf
809 Lundia: for Lund, Sweden, famous for its observatory
810 Atossa: name of sixth-century BC Persian queen, mother of
   Xerxes I
811 Nauheima: for German town of Nauheim (now usually
   known as Bad Nauheim)
812 Adele: identity unknown
813 Baumeia: for surname (Baum) of astronomer at Königstuhl
   Observatory, near Heidelberg
814 Tauris: apparently not for Latin name of Tabriz, Iran, but
   for ancient name (Taurida) of Crimea, where there was
   famous observatory (see 748 and 749)
815 Coppelia: for main character (doll created by Doctor
   Coppelius) in ballet by Delibes of same name
816 Juliana: identity unknown

817 Annika: identity unknown
818 Kapteynia: for Dutch astronomer, Jacobus Cornelius
       Kapteyn; see *Kapteyn's Star* in Dictionary
819 Barnardiana: for American astronomer, Edward Emerson
       Barnard (see 907 and *Barnard's Star* in Dictionary)
820 Adriana: identity unknown; perhaps first name of member
       of astronomer's family (Adrian), unless woman's name
       (Adriana)
821 Fanny: identity unknown
822 Lalage: identity unknown; perhaps for an assistant
       astronomer?
823 Sisigambis: mother of King Darius III (fourth century BC),
       taken prisoner by Alexander the Great together with
       Darius's wife and daughter
824 Anastasia: identity unknown, but doubtless a member of
       one of Russian astronomers' family (compare 825)
825 Tanina: identity unknown, but clearly refers to Russian
       name Tanya (short form of Tatyana or Tsvetana; compare
       785 and 824)
826 Henrika: identity unknown, but perhaps for Henrik rather
       than Henrike
827 Wolfiana: for German astronomer Max Wolf, of Heidelberg,
       discoverer of many minor planets
828 Lindemannia: said to be for Professor A.F. Lindemann,
       English astronomer
829 Academia: for Academy of Sciences, St Petersburg (now
       Leningrad, USSR; see 830)
830 Petropolitana: Latin-style name of St Petersburg (see 829
       and 857)
831 Stateira: Persian queen of fourth century BC, and wife of
       King Darius III
832 Karin: identity unknown; probably German
833 Monica: identity unknown; probably German
834 Burnhamia: for American astronomer, Sherburne W.
       Burnham
835 Olivia: identity unknown
836 Jole: daughter (Iole) of Eurytus, and later concubine of
       Heracles
837 Schwarzschilda: for Karl Schwarzschild, German
       astronomer, director of Potsdam Observatory
838 Seraphina: identity unknown

839 Valborg: name of heroine in drama *Axel and Valborg* by Danish poet Adam Oehlenschläger

840 Zenobia: widow and successor (in third century AD) of Odenathus, ruler of Palmyra, state in Syria, later captured and deposed by Aurelius

841 Arabella: heroine of Richard Strauss's opera of same name

842 Kerstin: unidentified

843 Nicolaia: for Torvald Nicolai Thiele, father of discoverer

844 Leontina: unidentified; perhaps member of astronomer's family

845 Naëma: unidentified

846 Lipperta: for Edward Lippert, founder of Hamburg Observatory

847 Agnia: for first name (Agny) of Russian scientist at Simeiz Observatory (see 748)

848 Inna: first name of Inna Balanovskaya, woman astronomer at Pulkovo Observatory (see 762)

849 *Ara*

850 Altona: for German city of Altona (now part of Hamburg)

851 Zeissia: for the German manufacturer of optical instruments, Carl Zeiss

852 Vladilena: for Vladimir Ilich Lenin, the Russian Communist leader; planet had been discovered (1916) at Simeiz (see 748) by S.I. Belyavsky, but was not named until 1924, year of Lenin's death

853 Nansenia: for Fridtjof Nansen, the Norwegian Arctic explorer

854 Frostia: for Edwin B. Frost, American astronomer and director of the Yerkes Observatory, Wisconsin

855 Newcombia: for Simon Newcomb, Canadian-born American astronomer

856 Backlunda: for Oskar A. Baklund, Russian director of Pulkovo Observatory (see 762)

857 Glasenappia: for Sergei P. Glazenap, Russian astronomer and director of Pulkovo and St Petersburg Observatories (see 762 and 830)

858 El Djezair: Arabic name of Algiers (see 859)

859 Bouzareah: district of Algiers where observatory is located (see 858)

860 Ursina: unidentified; perhaps operatic or literary character?

861 Aïda: heroine of Verdi's opera of same name (compare 871 and 978)

862 Franzia: for Julius H. Franz, German astronomer
863 Benkoela: apparently for Bengkulu (formerly Benkoelen), town in Sumatra, Indonesia, where presumably astronomical work was (is) carried out
864 Aase: character in Ibsen's drama *Peer Gynt*, in which she is hero's mother
865 Zubaida: wife of caliph of Baghdad, Harun al-Rashid, most famous of all the caliphs, and featuring in *Arabian Nights*
866 Fatme: daugher of founder of Islam, Muhammad (with her name better known as Fatima)
867 Kovacia: surname (Kovac) of doctor who cured wife of Austrian astronomer Palisa
868 Lova: unidentified
869 Mellena: for founder (von Melle) of Hamburg University
870 Manto: mythological soothsayer, daughter of Teiresias, a Theban seer
871 Amneris: King of Egypt's daughter in Verdi's opera *Aïda* (compare 861)
872 Holda: for Edward S. Holden, American astronomer at Lick Observatory, California
873 Mechthild: unidentified; German woman's name
874 Rotraut: unidentified; German woman's name
875 Nymphe: for one (or all) of mythological nymphs, spirits of nature living in streams, mountains, trees, and so on
876 Scott: for Robert F. Scott, English polar explorer
877 Walküre: for the valkyries, maidens of Norse mythology who served Odin and claimed corpses of dead warriors to take them to Valhalla (familiar from Wagner's operas) (compare 890)
878 Mildred: first name of daughter of American astronomer Harlow Shapley, director of Harvard Observatory (see 736); planet has since been lost
879 Ricarda: first name of German authoress Ricarda Huch
880 Herba: unidentified; mythological name?
881 Athene: best-known name of famous Greek goddess whose byname was *Pallas* and who was counterpart of Minerva (see 93); a late entry for her in the list!
882 Svetlana: unidentified, but clearly a Russian name; a literary character?
883 Matterania: surname (Matter) of a German manufacturer of photographic plates
884 Priamus: mythological King of Troy (also known as Priam)

885 Ulrike: unidentified, but perhaps for fortune-teller in Verdi's opera *A Masked Ball*

886 Washingtonia: for George Washington, first president of the USA

887 Alinda: unidentified; operatic character?

888 Parysatis: Persian queen of fourth century BC, daughter of Artaxerxes I

889 Erynia: one of Furies, collectively called Erinyes in Greek mythology

890 Waltraut: a valkyrie in Wagner's opera *The Twilight of the Gods* (compare 877)

891 Gunhild: unidentified; literary or operatic name?

892 Seeligeria: for German astronomer Hugo von Seeliger, director of Munich Observatory

893 Leopoldina: named for the 'Leopoldina' German Academy of Sciences (Deutsche Akademie der Naturforscher Leopoldina), the oldest German society of natural scientists, founded in 1652 and officially recognised in 1687 by Emperor Leopold I; from 1878 based at Halle, Germany (now East Germany)

894 Erda: Scandinavian nature goddess

895 Helio: perhaps for Helios, Greek god of the Sun

896 Sphinx: mythological female monster who asked riddles of young Theban men

897 Lysistrata: heroine of comedy of same name by Aristophanes

898 Hildegard: for the twelfth-century German saint and visionary

899 Jokaste: mother of Oedipus, who married her son without recognising him

900 Rosalinde: heroine (wife of Eisenstein) of Richard Strauss's opera *Der Rosenkavalier* (not 'name of an opera by Franz Schubert' as Paluzíe-Borrell states!)

901 Brunsia: for Ernest Bruns, director of Leipzig Observatory

902 Probitas: personification of probity, 'uprightness' (Latin)

903 Nealley: unidentified; presumably for an astronomer or his relation

904 Rockefellia: for John D. Rockefeller, American oil magnate and philanthropist, or for one of his descendants

905 Universitas: Latin for 'university'; name given for Hamburg University (founded 1919, in which year planet was discovered)

906 Repsolda: for Professor J.A. Repsold, precision engineer

907 Rhoda: first name of wife of American astronomer E.E.
    Barnard (see 819)

908 Buda: for Budapest, Hungarian capital (formed from former
    separate towns of Buda and Pest, either side of Danube)

909 Ulla: said to be name of woman acquaintance of discoverer

910 Anneliese: first name of wife of Max Beyer, Hamburg
    astronomer

911 Agamemnon: King of Mycenae who led Greeks at Troy and
    who was murdered by his wife Clytemnestra (see 179)
    and her lover Aegisthus

912 Maritima: for maritime expedition undertaken by members
    of Hamburg University

913 Otilia: name invented by German astronomer Karl
    Reinmuth without reference to a particular woman or
    girl (compare 799); for similar names see further examples
    below

914 Palisana: for Austrian astronomer Johann Palisa, vice-
    director of Vienna Observatory, and discoverer of over
    100 minor planets

915 Cosette: first name of youngest daughter of discoverer
    (Gonnessiat)

916 America: general compliment to astronomers and
    astronomical work of USA; discoverer was Russian
    astronomer Neujmin (compare 748 and 753)

917 Lyka: unidentified; name may be literary (?mythological or
    classical), and does not appear to fall into same category
    as ones below

918 Itha: name devised by Reinmuth (see 913), as are many
    below

919 Ilsebill: unidentified, and presumably in same category as
    918

920 Rogeria: name devised by Reinmuth, presumably based on
    Roger (see 913)

921 Jovita: name devised by Reinmuth, and perhaps based on
    'Jove' (see 913)

922 Schlutia: for acquaintance (Schlüter) of astronomer A.
    Kopff

923 Herluga: unidentified; may not be one of Reinmuth's
    devised names (see 940)

924 Toni: unidentified, but a recognised German name (see 940)

925 Alphonsina: said to be for two kings of Spain, both named

Alfonso; there were many: perhaps best known were thirteenth-century Alfonso el Sabio ('the Wise') and last of name (abdicated 1931, died 1941) Alfonso XIII

926 Imhilde: unidentified, but probably not one of Reinmuth's names (see 940)

927 Ratisbona: Latin name of Regensburg, German city where famous astronomer Kepler died (in 1630)

928 Hildrun: said to be name devised by Reinmuth, but does exist in own right

929 Algunde: apparently name devised by Reinmuth (see 913)

930 Westphalia: former Prussian province

931 Whittemora: for American professor Thomas Whittemore, of Universities of Harvard, Mass. and Columbia, New York

932 Hooveria: for Herbert C. Hoover, thirty-fifth president of USA; minor planet 1363 would subsequently (1935) be named Herberta, also for him

933 Susi: first name of wife of German astronomer K.R. Graff

934 Thuringia: former state of central Germany

935 Clivia: unidentified, but could be the flower of the name (itself named after Lady Charlotte Clive, Duchess of Northumberland)

936 Kunigunde: apparently for Cunégonde, heroine of Voltaire's novel *Candide*

937 Bethgea: said to be tribute to twentieth-century German novelist Hans Bethge

938 Chlosinde: unidentified

939 Isberga: unidentified

940 Kordula: unidentified; it is possible that these three names, and perhaps 923, 924, 926, 948, 963, 994 and 997 were taken from some *Lives of the Saints*

941 Murray: for British classical scholar Gilbert Murray, professor of poetry at Harvard and, more relevantly here, chairman of League of Nations (from 1923)

942 Romilda: unidentified; could be literary name or family member

943 Begonia: apparently for the flower

944 *Hidalgo*

945 Barcelona: for Spanish city, where planet was discovered (1920)

946 Poësia: personification of poetry

947 Monterosa: name of ship on which members of Hamburg

University made expedition, with naming taking place on board

948 Jucunda: unidentified; perhaps a saint's name (see 940)

949 Hel: alternative name of goddess Hela (see 699)

950 Ahrensa: for Ahrens family, who gave financial assistance to Königstuhl Observatory, near Heidelberg; family were friends of Reinmuth (see 913)

951 Gaspra: Russian resort and spa near Yalta, Crimea, where Tolstoy was treated; perhaps name given as tribute to him (compare 748)

952 Caia: unidentified; name may be literary

953 Painleva: for Paul Painlevé, French mathematician and statesman (twice premier of France in twentieth century)

954 Li: short form of first name (Lina) of wife of astronomer Reinmuth (see 913)

955 Alstede: maiden name of wife of Reinmuth (see 913)

956 Elisa: first name of mother of Reinmuth (see 913)

957 Camelia: apparently for flower of the name

958 Asplinda: for Danish astronomer, B. Asplind (see 959, 960 and 961)

959 Arne: first name of son of Asplind (see 958)

960 Birgit: first name of one of daughters of Asplind (see 958)

961 Gunnie: first name (short form of Gunhild) of second daughter of Asplind (see 958)

962 Aslög: unidentified, but appears to be Swedish name

963 Iduberga: unidentified; possibly an early saint's name (compare 940)

964 Subamara: unidentified; name (or word) happens to be Latin for 'rather bitter', but this may be irrelevant

965 Angelica: unidentified; perhaps for the plant, if not a personal name

966 Muschi: maiden name of wife of German-born American astronomer Walter Baade

967 Helionape: classical-seeming name is actually stage name of Austrian actor Adolf Sonnenthal (Greek *helios* and German *Sonne* mean 'sun'; Greek *nape* and German *Thal* mean 'valley' or 'glen'); name must have been tribute to him by discoverer

968 Petunia: for the flower

969 Leocadia: unidentified; based on surname?

970 Primula: for the flower

971 Alsatia: Latin name of Alsace, former province of north-east France

972 Cohnia: for German astronomer F. Cohn, who died in year of discovery (1922)

973 Aralia: for the decorative plant (member of ivy family)

974 Lioba: unidentified

975 Perseverantia: personification of perseverance

976 Benjamina: for first name (Benjamin) of discoverer's son

977 Philippa: for first name of Philip Rothschild, member of famous banking family, descendant of Albert Rothschild (see 719)

978 Aidamina: for Aida Minaevna, friend of discoverer, Russian astronomer S.I. Belyavsky, who chose this blended name as Aida had already been given (see 861; compare 979 and 981)

979 Ilsewa: for Ilse Waldorf, friend of the discoverer (see 978)

980 Anacostia: American place-name, either suburb of Washington or river that flows through it; perhaps used for city as actual name Washington already in use (see 886)

981 Martina: tribute to Cuban patriot and revolutionary, José Martí (killed by Spanish in 1895); name was given by Russian astronomer S.I. Belyavsky (compare 978)

982 Franklina: for Benjamin Franklin, American statesman and scientist

983 Gunila: unidentified

984 Gretia: said to have been devised by German astronomer Gustav Stracke without any individual reference (compare Reinmuth's similar names; see 913)

985 Rosina: unidentified

986 Amelia: unidentified; both this and previous name could have been for relatives of namer

987 Wallia: unidentified

988 Appella: for Paul-Émile Appell, French astronomer and president of many learned bodies, including Société Astronomique de France

989 Schwassmannia: for German astronomer, F. Schwassmann, of Hamburg University

990 Yerkes: for Yerkes Observatory, Wisconsin, where planet was discovered (1923) and for Charles T. Yerkes, who financed it

991 McDonalda: for McDonald Observatory, Fort Davis, Texas,

      where reflector telescope was constructed through financial aid from legacy of W.J. McDonald (compare 992)

992 Swasey: for optical engineer (Swasey) of Warner & Swasey Works, who constructed the reflector telescope at McDonald Observatory (see 991)

993 Moultona: for American astronomer, Forest R. Moulton

994 Otthild: unidentified; possibly early saint's name (compare 940)

995 Sternberga: for Russian astronomer, Pavel Karpovich Šternberg, director of Moscow Observatory

996 Hilaritas: personification of cheerfulness

997 Priska: unidentified; perhaps early saint's name (compare 940); not likely to be simply Latin *prisca* meaning 'ancient'

998 Bodea: for Johann E. Bode, German astronomer, director of Berlin Observatory

999 Zachia: for Franz Xavier von Zach, German astronomer, of observatory at Notre Dame des Anges, France (see 64)

1000 *Piazzia*

SELECT BIBLIOGRAPHY

Apart from standard English and foreign language dictionaries and encyclopaedias, the books and publications listed below were found to be the most helpful in determining the origin and meaning of the astronomical names in the Dictionary. Foreign titles are also translated for ease of reference.

Abbott, David (gen. ed.), *The Biographical Dictionary of Scientists: Astronomers*, Blond Educational, London, 1984.

Allen, Richard Hinckley, *Star Names: Their Lore and Meaning*, Dover Publications, New York, 1963 (originally published by G.E. Stechert in 1899 as *Star-Names and Their Meanings*).

Andersson, Leif E. and Whitaker, Ewen A., *NASA Catalogue of Lunar Nomenclature* (NASA Reference Publication 1097), National Aeronautics and Space Administration, Washington DC, 1982.

Audouze, Jean and Israël, Guy (eds), *The Cambridge Atlas of Astronomy*, Cambridge University Press/Newnes, Cambridge/London, 1985.

Blagg, Mary A. and Müller, K., *Named Lunar Formations*, Percy Lund, Humphries, London, 1935.

Coulet du Gard, René, 'The Arabs are Coming: An Onomastic Alchemy', *Onomastica* (Journal of the Canadian Institute of Onomastic Sciences), 54, 1978.

Efemeridy malykh planet na 1987 god ('Ephemerides of the Minor Planets for 1987'), Institute of Theoretical Astronomy, Academy of Sciences of USSR, Leningrad, 1986.

Evershed, Mrs M.A., 'Who's Who in the Moon', *Memoirs of the British Astronomical Association*, 34, Neill, Edinburgh, 1943.

Herget, Paul, *The Names of the Minor Planets*, Cincinnati Observatory, Cincinnati, 1955.

Heuter, Gwyneth, 'Star Names – Origins and Misconceptions', *Vistas in Astronomy*, 29, 237–51, Pergamon Journals, Oxford, 1986.

Ideler, Ludewig, *Untersuchungen über den Ursprung und die Bedeutung der Sternnamen* ('Enquiries into the Origin and Meaning of Star Names'), Johann Friedrich Weiss, Berlin, 1809.

Karpenko, Yu. A., *Nazvaniya zvëzdnogo neba* ('Names of the Starry Sky'), Nauka, Moscow, 1981.

Kenny, Hamill, 'Place-Names on the Moon: A Report', *Names* (Journal of the American Name Society), 12, 2, 1964.

Kunitzsch, Paul, *Arabische Sternnamen in Europa* ('Arabian Star Names in Europe'), Otto Harassowitz, Wiesbaden, 1959.

Lipsky, Yu. N., *Atlas obratnoy storony Luny* ('Atlas of the Far Side of the Moon'), pt 3, Nauka, Moscow, 1975.

Menzel, D.H., Minnaert, M., Levin, B., Dollfus, A., Bell, B., 'Report on Lunar Nomenclature by the Working Group of Commission 17 of the IAU', *Space Science Reviews* 12 (1971), D. Reidel, Dordrecht, 1971.

Mitton, Jacqueline and Mitton, Simon (eds and introd.), *Star Atlas*, Cape, London, 1979.

Mitton, Simon (ed.), *The Cambridge Encyclopaedia of Astronomy*, Cape, London, 1977.

Moore, Patrick, *Patrick Moore's A–Z of Astronomy*, Patrick Stephens, Wellingborough, 1986.

Moore, Patrick, *The Guinness Book of Astronomy Facts and Feats*, Guinness Superlatives, Enfield, 2nd edn, 1983.

Moore, Patrick, *The Atlas of the Universe*, Mitchell Beazley, London, completely revised and updated edn, 1981.

Moore, Patrick and Hunt, Garry, *The Atlas of the Solar System*, Mitchell Beazley/Royal Astronomical Society, London, 1983.

Murdin, Paul and Allen, David, *Catalogue of the Universe*, Cambridge University Press, Cambridge, 1979.

Norton, Arthur P., *Norton's Star Atlas and Reference Handbook*, ed. by Gilbert G. Satterthwaite in consultation with Patrick Moore and Robert G. Inglis, Gall & Inglis, Edinburgh, 17th edn, 1978.

Paluzíe-Borrell, Antonio, *The Names of the Minor Planets*, Jean Meeus, Kessel-Lo, 1963.

Ridpath, Ian, *Guide to Stars and Planets*, Collins, London, 1984.

Ridpath, Ian (ed.), *The Illustrated Encyclopaedia of Astronomy and Space*, Macmillan, London/Basingstoke, rev. edn, 1979.

Rükl, Antonín, *Maps of Lunar Hemisphere*, D. Reidel, Dordrecht, 1972.

Rükl, Antonín, *Moon, Mars and Venus*, Hamlyn, London, 1976.

Simon, Tony, *The Search for Planet X*, Scholastic Book Services, New York, 1962.

Tombaugh, Clyde W., *Out of the Darkness: The Planet Pluto*, Lutterworth, Cambridge, 1980.

Webb, E.J., *The Names of the Stars*, Nisbet, London, 1952.

Wilkins, H.P. and Moore, Patrick, *The Moon*, Faber, London, 2nd edn, 1961.